PLANTWATCHING

PLANTWATCHING

How Plants Live, Feel and Work

MALCOLM WILKINS

MACMILLAN
LONDON

To Pat and Nigel
and in memory of
Fiona

First published 1988 by
MACMILLAN LONDON LIMITED
4 Little Essex Street London WC2R 3LF
and Basingstoke

Associated companies in Auckland, Delhi, Dublin, Gaborone, Hamburg, Harare,
Hong Kong, Johannesburg, Kuala Lumpur, Lagos, Manzini, Melbourne, Mexico City,
Nairobi, New York, Singapore and Tokyo

This book was created and produced by Roxby Reference Books Limited
a division of Roxby Press
126 Victoria Rise
London SW4 0NW

Editor: Martyn Bramwell
Design: Eric Drewery
Picture Research: Damien Wilkins
Typesetting: Hobbs the Printers Limited, Southampton
Reproduction: J. Film Process Co. Limited

British Library Cataloguing in Publication Data
Wilkins, Malcolm
Plantwatching: how plants live, feel and work
1. Botany – Ecology
I. Title
581.5 QK901

ISBN 0-333-44503-1

Printed by New Interlitho, Milan

CONTENTS

CHAPTER 1

Plants in Perspective

Plants inhabit almost every part of the world, from the open seas to the heat and humidity of the tropics and the intense cold of the polar ice caps. Some are microscopic, single-celled organisms; others are enormous structures – the largest living organisms the world has ever seen. Except for the fungi and a few parasitic species, all plants have a unique characteristic shared by no other living organisms except one group of bacteria, and that is the ability to capture and store the energy in the sun's rays. It is this characteristic that makes them so important, for they, and they alone, provide the food-energy resources on which all other living things depend. Without the green plant and its capacity to carry out the process of photosynthesis, the earth would be a dead planet.

The indispensable plant

Mankind is absolutely dependent on natural vegetation in primitive societies, and on organized agriculture, horticulture and forestry in advanced societies, for food and many other necessities of life. The level of this dependence is all too clear in the least developed countries of the world where widespread suffering and death occur when natural vegetation is damaged or when crops fail. Such disasters may be caused by drought, or by flooding or by epidemics of plant diseases, but today they are increasingly precipitated by man's own actions – by war, pollution, inadequate agricultural practices and by simply overloading the land with people and livestock.

Green plants are the sole source of food for all animal life on earth, including mankind, and the feeding of the world's burgeoning population depends entirely on our ability to produce enough food crops for ourselves and enough fodder for our domestic livestock. The future success of these efforts will require the use of ever more marginal land, the development of much more efficient growing, harvesting and food-storing techniques, and perhaps above all the further development of new strains of crop plants capable of withstanding hostile conditions such as prolonged drought, poor or saline soils, and the ravages of pests and diseases.

Over and above the technical problems of increasing food productivity is the fundamental problem that there is only a finite amount of usable land on the earth's surface, and that the human population is now growing so fast that it is bound, eventually, to exceed the maximum productive capacity of the land. The global suffering and hardship that such a situation would cause may still be a century or more away, but the impact it would have can already be seen at local and regional level in many parts of the world. The world's population is now growing by a quarter of a million people every day, and while the brilliant advances made by plant breeders and geneticists in developing new high-yielding strains of wheat and rice have transformed food production in many countries, the gains achieved through such advances will be wiped out within a few decades if we fail to bring population growth under control.

Plants, however, do far more than just provide food for mankind and his livestock. Nearly half of the world's population – some 2,000 million people – depend entirely on wood to

This view up through the leaf canopy of a cloud forest at an altitude of 1,000 metres in Rancho Grande shows how effective plant foliage is in absorbing the sun's radiant energy.

heat their homes and cook their food, while the rest of the world derives its industrial and domestic energy needs from the fossilized remains of forests and swamps that clothed the earth 300 million years ago. Around the world, logs and sawn timber, bundled reeds and woven leaves are the universal building materials for homes. Softwood plantations in the temperate regions feed building and packaging industries and paper mills, while the natural forests of the tropics provide exotic hardwoods like mahogany and teak which are so prized for quality furniture. Perhaps least understood of all are the chemical resources of the plant kingdom – especially the genetic and pharmaceutical resources of the world's tropical forests. The genetic resource could hold the key to the development of many new high-yield or disease-resistant food plants, yet the forests are being destroyed so fast today that hundreds, perhaps thousands, of plants are likely to become extinct before they are even named, let alone studied. And in the field of medicine the losses are just as great. Despite dramatic advances in synthesizing new drugs, about half those in use today are still extracted from plants. They include the opiate drugs produced by the poppy, the heart stimulant digitonin from the foxglove, and from the tiny rosy periwinkle of Madagascar one of the most effective compounds yet found for the treatment of leukemia in children. Plants are without doubt the best synthetic chemists on earth. They can manufacture countless complex organic compounds that man as yet has no idea how to make in the laboratory. We have barely begun to tap the vast potential of this resource.

The versatility of plants

The fact that plants are found in all parts of the world shows that they are able to tolerate an astonishing range of climatic conditions. Of course, a plant can not be transplanted from a tropical region to a polar region, or from a rainforest to a desert, and be expected to survive, because each will have developed special features that suit it to its normal climatic conditions. But the fact that plants live successfully in hot and cold climates, in the driest deserts and submerged in water shows that they have been able to evolve and adapt in a great variety of ways, with the result that some plant or other can tolerate just about any combination of environmental conditions our planet has to offer.

In order to tolerate the new environmental conditions they encountered as they spread out to cover the earth's surface, plants had to adapt and change not only their physical structure but also the biochemical and physiological machinery by which they function. For instance, when plants first emerged from the sea or freshwater lakes to colonize the land, they left behind a watery environment that supported them, insulated them from extremes of temperature, and provided them with all the inorganic (mineral) nutrients they required. At the same time it allowed them to absorb the sun's rays and, by photosynthesis, make their own carbohyrates (sugars and starch) from the carbon dioxide dissolved in the water. The water also provided them with a liquid medium in which their delicate, free-swimming reproductive cells could easily swim and fuse, so producing the fertilized cells of sexual reproduction from which the next generation would develop.

Once on land, plants were faced with a staggering number of problems that simply do not exist in the sea or in a freshwater lake, and all of them had to be solved if the plants were to have any chance of survival. First of all the leaves needed some form of mechanical support to hold them up to the rays of the sun. Then the surface of the plant had to be protected against excessive water loss, otherwise the plant would dry out, but this had to be achieved in such a way that important gases such as oxygen and carbon dioxide could still pass in and out of the plant without undue hindrance. Once out of the water, some means was needed of protecting the delicate reproductive cells, and of transferring them from one plant to another without damage so that sexual reproduction could take place. Mechanisms were also needed to enable the plant to tolerate the widely different climatic conditions that occur at different times of the year, and once removed from its 'nutrient bath' the plant required an internal transportation system to carry water and nutrients from one part of its structure to another. Finally, a way had to be found of ensuring that the plants could reproduce themselves at a time of year when their offspring would have the best chance of survival, and of providing the vulnerable young plants with physical protection and with stored food reserves to nourish them until they became self-supporting.

Successful solution of these problems has involved major changes in the plants' structure and metabolism, that is, their biochemical machinery, as well as the acquisition of sophisticated sensory mechanisms without which they would have no chance of survival on land. To take just one example, in northern temperate regions, deciduous perennial plants stop producing normal green foliage leaves and begin producing winter buds in August and September, long before the cold winter weather begins. This shows that such plants do not just 'live for the moment', but can anticipate the approaching adverse conditions and prepare for them well in advance. Being able to predict the future in this way seems, at first sight, to belong to the realm of mysticism, but this is not so; it involves the operation, in the plant, of a mechanism for measuring the one environmental factor that changes in an absolutely regular way each year – and that is the length of the night. You may wonder what is so remarkable about that, but pause for a moment and ponder how *you* would measure the length of the night to an accuracy of, let us say, five or ten minutes. It is certain that the answer would be by using a clock or watch; in fact, there is no other way of doing it. A plant that can measure the length of the night to this degree of accuracy must, therefore, be in possession of a time-measuring mechanism or 'biological clock'. This mechanism will be discussed in detail in a later chapter but its existence illustrates the point that even the simple observation of when plants begin to make their winter buds can very quickly lead us to realize just how sophisticated and fascinating plants can be.

This book sets out to describe and explain how plants live, work and behave. All plants, like animals, live in a competitive world and the most successful are those that grow quickly into the most efficient structures for capturing the sun's energy, and that compete most effectively for the mineral nutrients and water in the soil. Their growth, development, reproduction and survival require complex and finely tuned control and sensory systems to ensure that effort is not wasted – for example, by an organ such as a root or stem growing in the wrong direction, or by flowers appearing in the wrong place or at the wrong time.

We are naturally more familiar with the sensory systems in animals than with those in plants, but plants have developed equally sophisticated systems. While plants apparently lack the capacity to communicate with one another by sound, they have, for example, at least three different light-sensing systems, each of which involves a different light-absorbing mechanism and controls an entirely different set of functions. By contrast, most animals have only a single light-sensing mechanism – the eye. Apart from the exception mentioned above, plants can do almost everything animals can do, but usually rather more slowly, and this is one reason why time-lapse filming techniques have been so successful in awakening popular interest in plant behaviour.

Plants grow and develop in highly organized ways. They generate enormous forces, which can destroy roads, buildings and pavements; they constantly sense their environmental conditions of light, temperature and gravity; they measure time; some of them can count, some have a memory and some have a sense of touch; they feed, respire and absorb nutrients in a selective way from the soil; they recognize one another when brought into physical contact, and in many cases they move about. They also have systems to combat infection, while many are able to enter into mutually advantageous relationships with one another and with certain kinds of bacteria and fungi. In addition, they have the remarkable property of being able to regenerate themselves from their smallest components – single cells – even though these cells may be taken from a highly specialized organ such as a root or petal. What this means is that a cell taken, for example, from a root must contain all the information for making leaves, stems and flowers, even though this information was not being utilized in the root. In other words, every living plant cell contains a complete genetic 'blue-print' of the whole plant. How else could a gardener expect a stem cutting to be able to produce the roots he requires for the successful establishment of a new plant?

In attempting to explain how plants do all these things, we must admit at the outset that substantial areas of plant function and behaviour are not understood very well, and that much remains to be investigated. For the most part, this book will concentrate on the behaviour of the most advanced group of plants, the angiosperms or flowering plants, since it is to this group that all the food crops and most other commercially important plants belong and, understandably, it is on the crop plants and their close relatives that governments, universities and industries have concentrated most of their research.

CHAPTER 2

Building Bricks and the Generation Gap

In order to appreciate how plants live, grow, develop and reproduce, we must first understand the basic units, or 'building bricks', from which they are made. These building bricks are called cells, but unlike the bricks that go into the construction of a house they are not all the same: they exist in a wide variety of shapes and forms, each specialized to carry out a certain function or task. All these different cells develop from identical young cells by a process called cell differentiation, a highly regulated process in which specialized chemical reactions control each cell's length and width, the thickness and architecture of its walls, and the nature of its contents. Since the growth of a whole plant involves the development and growth of new roots, leaves, stems, flowers and other structures, there clearly has to be a continuous supply of new building bricks to add to those already in existence. This supply is achieved by a process called cell division.

Within the plant there are localized areas of cell division. The principal ones are called meristems, and the most obvious are those at the tips of the growing roots and stems. These are called apical meristems. Other important areas of cell division develop later in the life of the plant when the stems and roots of both annual and perennial plants begin to thicken. This process of secondary thickening is described in more detail in Chapter 10 because it is associated with the provision of additional pipework for the plumbing of the plant as it increases in size.

In this chapter we shall first be concerned with what the young plant cells in an apical meristem look like and how they divide to produce the continuous supply of new cells needed by the growing plant. Then we shall consider the differences between body cells and reproductive cells which, for reasons that will become clear, have to be very different from normal body cells.

The plant cell

A newly formed plant cell is a more or less rectangular structure consisting of a number of distinct parts. Each part is made up of different types of chemical substances, and each is designed to perform a different task or series of tasks essential to the well-being of the healthy plant.

The outermost layer is called the cell wall and is made up almost entirely of long chains, or polymers, of sugar molecules – the principal one being the glucose polymer called cellulose.

Even the modest power of an optical microscope (*right*) will reveal the variety of cell types and structural features in the stem of a buttercup.

The enormous resolving power of a scanning electron microscope (*below*) takes us into another dimension altogether. Here, inside a cell of the green alga *Chara*, we can see the surface of the nucleus itself—the control centre of the living cell.

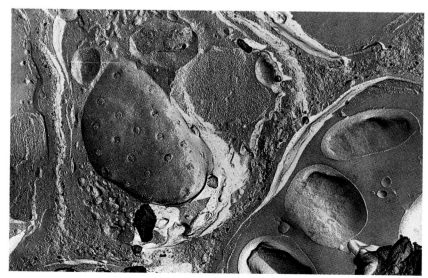

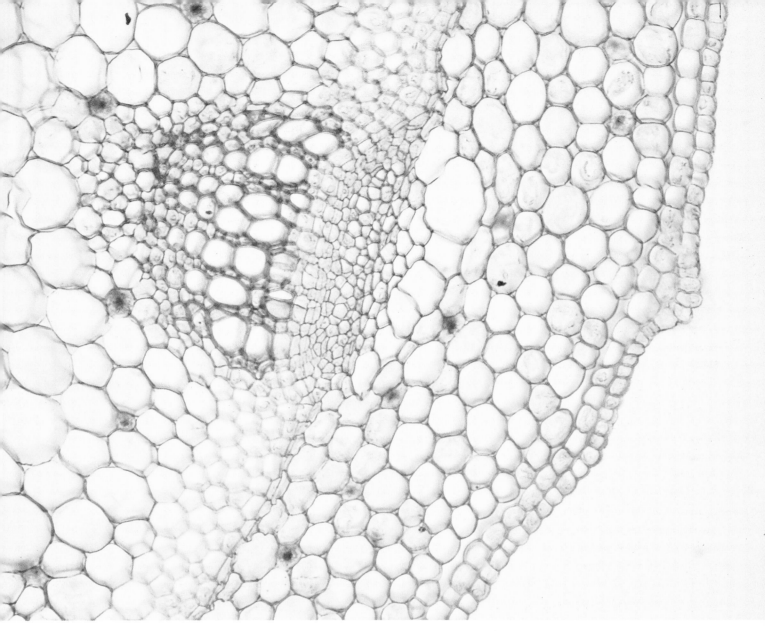

Viewed under an electron microscope, the anatomy of a plant cell is laid out like a map. Nucleus, mitochondria and chloroplasts float in the jelly-like cytoplasm, contained by the plasmalemma and protected by the tough cellulose cell wall.

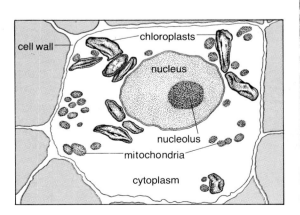

cell wall

chloroplasts

nucleus

nucleolus

mitochondria

cytoplasm

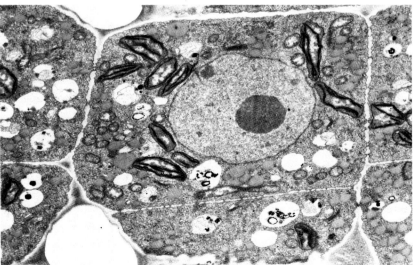

The cellulose molecules are very long and thin and they are grouped together in bundles to form strands called microfibrils which form a multi-layered mat-like structure of great strength around the cell. The cell wall is rather like the outer leather case of a football, except that its open texture allows water and any dissolved substances an entirely free passage.

Within the cell wall there is a jelly-like mass called the protoplasm, which is divided into two main parts – the nucleus and the cytoplasm. The nucleus is essentially an information store. It holds the detailed instructions needed by the plant cells to manufacture the great variety of chemicals they will require throughout their life, and also for the formation of the various specialized types of cell produced in the process of cell development. The nucleus also contains instructions for the formation of complex organs such as roots, leaves, stems, buds and flowers – all of which require cell division and growth to proceed along precise and predetermined courses from which little or no deviation can be tolerated. After all, how could two orchid flowers be so alike unless the development of each had followed exactly the same set of instructions?

The cytoplasm is the factory of the cell, where a huge number and variety of chemical reactions take place. Some break down compounds like sugars and fats, releasing the energy they contain so that it can be used to drive other chemical reactions, which in turn produce the many chemicals required for the growth and development of the cell. The cytoplasm is a very complex semi-liquid substance in which are embedded a number of other bodies, each with a distinct shape and function. It is surrounded by a very important structure called the cell membrane, or plasmalemma, which separates it from the cell wall.

This cell membrane is extremely thin – just two molecules thick. It consists of phospholipids – molecules comprising a lipid or fat molecule on to which is joined a chemical group containing phosphorus. Embedded in this phospholipid layer are large globular protein molecules which pass right through the membrane and protrude from both the inner and outer surfaces. The importance of this thin and very delicate membrane is that unlike the cell wall it is *not* freely permeable to all types of chemical molecules. Indeed, it can best be regarded as being freely permeable to only one type of molecule – water. It is called a

semipermeable membrane, and it is this important property that enables the plant not only to stand erect but also to exert enormous forces on its surroundings, as will be described in Chapter 6. Molecules other than water do, however, pass through the membrane, both inwards and outwards, but they do so only by invitation. They are actively transported, or pumped, inwards or outwards across the membrane by very special mechanisms which first identify their chemical structures and then, if these are 'approved', transfer the molecules across the membrane in the required direction. These molecular transport mechanisms are associated with the large proteins embedded in the membrane, and they provide a means whereby plant cells can absorb particular chemical substances from the soil or from the sea, and accumulate them, while entirely rejecting others which they do not require or which may even be harmful.

Within the cytoplasm there are, in addition to the nucleus, a number of other bodies of various shapes and sizes. Each of these is surrounded by a membrane virtually identical to the plasmalemma. Two of the most important of these sub-cellular particles are called mitochondria and ribosomes. The mitochondria are oval or sausage-shaped structures which contain much of the chemical machinery concerned with the process of respiration – the breakdown of sugar which releases energy for other purposes (Chapter 12). The ribosomes are rather smaller particles which contain the chemical machinery for assembling amino acid molecules into long chains called proteins. When made to certain very specific patterns, many of these proteins have the property of causing particular chemical reactions to take place. They are called enzymes. In addition, in the cytoplasm there are sheets of membrane called endoplasmic reticulum. These appear to play a role in communication within and between cells since they seem to be a continuation of the membrane surrounding the nucleus and are interconnected with membranes which pass through the cell walls from one cell to the next.

In the central part of a fully mature plant cell there is usually a large vesicle, or cavity, called the vacuole which contains a solution of many chemical substances and inorganic ions. It is used by the plant cell as a reservoir for storing water and many of the chemical substances it manufactures and absorbs.

Finally, in plant cells, there are structures called plastids. These, too, are surrounded by membranes, and may develop very complicated internal structures when they are mature. The chloroplast is the main one for it is this body that contains the chlorophyll and other pigments which alone capture and store the sun's radiant energy. In the young cell these are not yet fully developed, but the bodies from which they develop are present. Chloroplasts will be discussed in some detail in Chapter 9.

The plant cell is, therefore, a very complex structure which is best likened to a football in which the rubber bladder has been filled with a very thin jelly containing a number of different-sized granules. The outer leather case of the football is equivalent to the cell wall, the rubber bladder to the cell membrane, and the jelly-like contents to the protoplasm in which are embedded the nucleus, mitochondria, ribosomes, plastids, the endoplasmic reticulum and a number of other structures. The vacuole would be like another smaller bladder, filled with water and embedded in the jelly-like contents of the main bladder. The outer wall gives the cell considerable mechanical strength, especially since the cell contents become pressurized (Chapter 6) like an inflated football.

Cell division

The continuous supply of cells needed for plant growth comes from a process of cell division. This process takes place especially in the apical meristems of the root and shoot and involves a rather important series of steps. If a cell simply divided into two it is likely that each new cell would, by chance alone, contain about half the cytoplasm and about half of the plastids, mitochondria, ribosomes and endoplasmic reticulum since there are hundreds of these structures in each cell and they are fairly evenly distributed throughout the cytoplasm.

What each new cell would *not* contain, however, would be a nucleus, because the original cell contains only one, and this would have to go into one or other of the new daughter cells. One cell would therefore end up without the basic library of information it requires for its growth and development. It is necessary therefore that before the moment of cell division occurs, a process takes place that leads to the division of the *nucleus*. The nucleus can not, however, just divide randomly into two bits for there would be no guarantee that each bit would contain a complete set of the basic instructions contained in the original. The division of the nucleus therefore has to be a very precise process to ensure that each daughter cell receives a full set of basic information. This division process is called mitosis, and to appreciate how it takes place we must first understand how the information is stored in the nucleus.

The information in a cell nucleus is stored on structures called chromosomes. These are long threads of a chemical called deoxyribose-nucleic acid, or DNA, and particular bits of information are held in short sections of the threads called genes; that is why it is called genetic information. There may be hundreds of different genes on a single chromosome thread, and the nucleus of each cell has a fixed number of chromosomes that make up a complete set. Each species of plant has a specific number of chromosomes in its complete set, and the number can vary from as few as 4 to the more usual number of between 30 and 45. In wheat, for example, there are 42 chromosomes, though in some ferns there can be as many as 260.

If all the information on each chromosome is to be passed on to each of the two daughter cells that result from a cell division, it is clear that each chromosome must duplicate itself longitudinally, divide longitudinally, and the two halves separate and travel to opposite ends of the original cell before the cell itself divides into two. This is just what is achieved in the process of mitosis, and it can be observed under the microscope.

When the process of division begins, the normally-invisible thread-like chromosomes become visible in the nucleus. This is because they shorten dramatically and become much thicker. During the time between the last division and the one about to begin, the chromosomes have, in fact, made an exact copy of themselves along their whole length so that when the chromosomes appear in the nucleus in the early stages of a division they each consist of two identical threads called chromatids, joined together at a point along their length called a centromere. The chromatids then become very short and thick, and the nuclear membrane around them disappears, leaving them 'floating' loose in the cytoplasm. They arrange themselves along the central part of the cell and then each chromosome separates completely into its two chromatids. One chromatid from each pair moves to each end of the cell along a fibrous structure called a spindle. The sets of chromatids then become enveloped in new nuclear membranes, thus forming the nuclei of the two new cells. Each chromatid now becomes a chromosome, extending to its original length and becoming invisible. Once this nuclear division or mitosis is complete, the cytoplasm divides along the central region of the cell where the chromosomes originally began their separation process. The cytoplasm of the two new cells is first separated by two new cell membranes, which develop side by side and join up with that surrounding the original cell. Then a cellulose cell wall is secreted by the cytoplasm of each new cell.

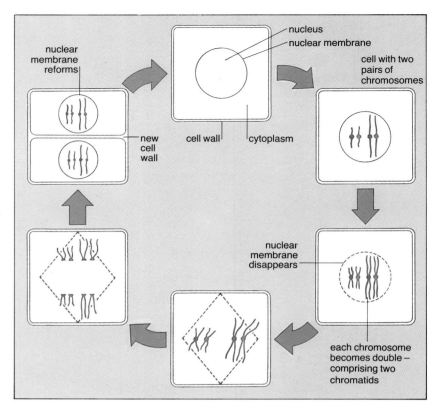

The process of cell division is now complete, and each new cell has a full complement of nuclear genetic information and a large number of the other cytoplasmic bodies.

The continued division of plant cells in the meristems of roots and shoots therefore produces large numbers of identical cells, each having the same number of chromosomes as the original, and hence a full set of genetic information. The cells subsequently differentiate into a variety of different types, many of which will be described in later chapters when their particular functions are discussed.

Reproductive cells or gametes

Just as the chromosomes in the nucleus have to reproduce themselves exactly in order to ensure that a complete set of genetic information is passed on to each of the new cells produced by mitotic division, so a similar kind of problem arises when reproductive cells, or gametes, are produced.

Gametes – that is, sperm and egg cells – fuse together to form a zygote at the moment of fertilization. It follows, then, that without some additional procedure each zygote would receive *two* complete sets of genetic information. The number of chromosome sets in each succeeding generation would therefore double – which

Mitosis is the process of nuclear division by which new cells are made so that a plant can grow. Inside the nucleus, each member of a pair of chromosomes duplicates itself along its length, forming a pair of chromatids. These then separate and migrate to opposite ends of the cell where they are enclosed in new cell membranes and then by new cell walls. In this way each of the two daughter cells has the same number of chromosomes as the original cell.

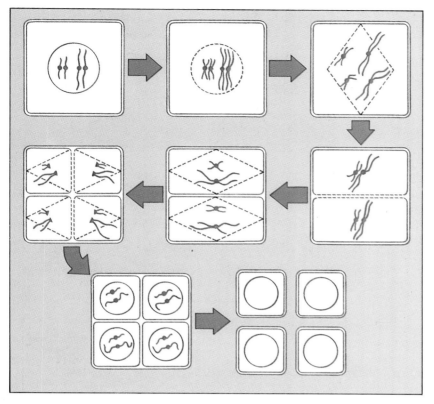

Meiosis is the special form of cell division that produces the eggs and sperm cells needed for sexual reproduction. As these cells combine at the moment of fertilization, each must carry only half the normal complement of chromosomes, and this is achieved by a two-stage process that produces not two but four daughter cells, each containing half as many chromosomes as the original cell.

would be catastrophic after even a relatively small number of generations! Careful examination of the number of sets of chromosomes in the body cells of succeeding generations shows that it remains constant. In the formation of gametes, therefore, there must be a special procedure which effectively reduces by half the number of sets of chromosomes that enters the nucleus of each gamete. This reduction is achieved in a special type of nuclear division called meiosis.

The body cells in most familiar animals and higher plants contain not one set of chromosomes in their nuclei, but two. These cells are called diploid cells, and it is only in such cells that meiosis can take place. What happens in meiosis is that the diploid cell undergoes *two* successive nuclear and cell divisions which result in *four* daughter cells, each of which has only *one* set of chromosomes. Such cells are said to be haploid. The first division is called the reduction division because one set of chromosomes passes into each of the two new cells, while the second division is identical with mitosis, with each new cell dividing and passing on to its daughter cells one complete set of chromosomes.

In the first or reduction division the two sets of chromosomes appear as ever-thickening,

ever-shortening threads in the nucleus. Each chromosome has already duplicated itself longitudinally to consist of two chromatids. Corresponding members of the two sets of chromosomes come to lie side by side and the four chromatids often exchange parts of their length with one another. The pairs of chromosomes arrange themselves in the centre of the cell and then one whole chromosome of each set (i.e. the two chromatids joined at their centromere) moves to each end of the cell. The cytoplasm then divides and so two cells are produced each now having only one set of chromosomes. The nuclear membrane does not yet reappear but the single set of chromosomes in each cell now arranges itself along a line in the new cells at right angles to the first division. The centromeres of each chromosome divide and the chromatids move away from each other to form new nuclei. The cytoplasm then divides, the nuclear membranes reappear and cell walls are formed so that four new cells are produced, often referred to as a tetrad. Each of these cells has only one set of chromosomes and is therefore haploid. These cells may immediately develop into gametes, as they do in animals, but in plants they usually undergo further and often extensive mitotic division to produce a special structure (or even a separate plant!) before gametes are formed. This gives rise to an important phenomenon known as the alternation of generations.

Alternation of generations; the biological 'generation gap'

In animals, when a diploid body cell undergoes meiosis in the sex organs, the four haploid cells that are produced become gametes – either eggs or sperms. These are the only haploid cells produced by the animal, and they develop no further unless they meet and fuse with their opposite number to form a zygote. This fusion of two haploid cells restores the diploid condition of having two sets of chromosomes and the zygote immediately begins to develop into a new animal.

In plants, the situation is much more complicated in that meiosis does not lead directly to the formation of gametes. Instead the newly-formed haploid cells may continue to divide, and in this way they grow and develop into distinctive structures which can vary in size and complexity from one group of plants to another. In the ferns, for example, quite separate plants develop consisting entirely of haploid

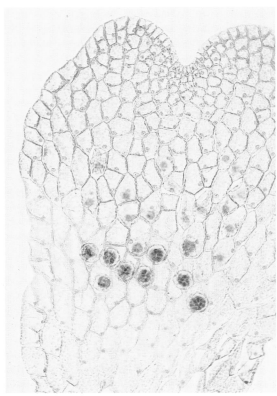

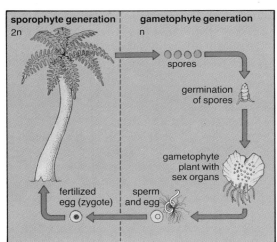

The trunk-like stems of some tree ferns may grow as high as 25 m. This stand in Papua New Guinea (*left*) consists of sporophytes of the tree fern *Cyathea*.

The gametophyte plant of the fern *Dryopteris* (*top of page*) is no more than 1cm long. The female reproductive organs (archegonia) are the dark spots near the apex of the plant. The even darker male organs (antheridia) can be seen lower down.

The diagram above shows the main stages in the life cycle of a tree fern. The dominant plant, the sporophyte, has two sets of chromosomes in each cell (2n). The very much smaller gametophyte plant is haploid, and produces the single-set (n) eggs and sperm that will fuse to make a new sporophyte.

cells. In shape and appearance this plant is quite different from the one containing only diploid cells, despite the fact that both are members of exactly the same species. The diploid plant is called a sporophyte generation because it produces spores. These are cells which develop directly from the four tetrad cells produced by meiosis, and they are haploid. They usually have a thick protective cell wall and can withstand being dried. They are released from the diploid plant and when they arrive at a suitable place they start to divide by mitosis to produce the haploid plant. This new plant is called the gametophyte generation because it will eventually produce gametes. So a spore is a haploid cell that can divide, grow and develop into a gametophyte generation without having to fuse with any other cell. It is not a gamete, neither is it a seed, for it contains no embryo and has not resulted from sexual reproduction!

The haploid gametophyte plant of a fern grows and develops to maturity at which point it produces relatively simple sex organs in which quite distinctive reproductive cells called gametes develop. These vary in shape and size from one group of plants to another, but in most non-flowering plants the egg cells are contained in a simple flask-shaped structure called an archegonium and the sperm in a large spherical structure called an antheridium. In most of these lower plant groups – that is, the algae, mosses and ferns – the sperm cells are able to move about under their own power. They have a number of thread-like flagellae with which they can swim as long as there is a film of water present, and in this way they reach and enter the archegonium to fertilize the egg. Without first fusing together, neither the egg nor the sperm can develop any further because both cells are haploid. Once the egg has been fertilized, however, it becomes a zygote. It now has two sets of chromosomes, one from the egg and one from the sperm, and so it becomes the first cell of a new sporophyte generation.

Because, in all groups of plants, gametophyte generations alternate with sporophyte generations, the life cycles and life-styles of plants can be very varied. In the more primitive land plants, the gametophyte is usually the larger and dominant generation with the sporophyte being in some cases almost completely dependent on it for support and nutrition, as in the case of the mosses and liverworts. In the large

group of plants to which the ferns belong, the gametophyte and sporophyte plants are quite independent, and each can live for many years. In some members of this group there are, in fact, two quite distinct gametophyte plants, one male and the other female, as well as a sporophyte plant! In the more advanced plants like the conifers, and ultimately in the flowering plants, the sporophyte is the dominant generation and the gametophyte is much reduced in size, dramatically so in the case of flowering plants where the male gametophyte consists of just one or two cells within the pollen grain.

It seems clear that in the course of plant evolution the sporophyte has had a much greater capacity to develop into a complex plant structure able to withstand the rigours of life on land. The reduction in size of the male gametophyte to only one or two cells contained within the protective coat of the pollen grain has been an essential step in enabling plants to colonize the land environment successfully. Free-living gametophytes appear never to have been able to develop into structures capable of escaping from a wet and humid environment. Furthermore, wet and humid conditions are required if the vulnerable sperm are to reach and fuse with egg cells, and this creates another barrier to land colonization.

The evolutionary changes in the relative importance of the gametophyte and sporophyte generations, and the abolition of free-swimming sperm and dangerously exposed egg cells in favour of protected gametes and a robust sporophyte, will be highlighted in Chapters 3 and 4 on the plant kingdom and its varied life-styles.

The gametophyte generation of the liverwort *Pellia* is a simple flat structure called a thallus, which supports and also nourishes the tiny sporophyte plant. This consists of a foot, embedded in the thallus, a short white stalk, and a shiny black sporangium which contains the microscopic spores.

CHAPTER 3

Variety in Life-styles: The Seedless Plants

The plant kingdom today comprises six groups of plants with widely differing structural and reproductive features, and in comparing them it is possible to see how plants have changed in order to adapt to conditions on land. However, although the groups are arranged and described in this and the next chapter in order of increasing complexity, it should not be assumed that one has evolved from another. Plant evolution is a matter of considerable speculation, and it is likely that several groups developed from a common ancestor among the green algae and subsequently followed several different lines of evolution. There is a fascinating fossil record of plants and it is from these fossils that deductions are made about the evolution of the different present-day groups. In this book we will consider fossil plants only in relation to the ferns and their allies, since it is in this group that the first real land plants probably arose.

The four major groups of seedless plants are the algae – seaweeds, and very simple green plants found in fresh water; the fungi – non-photosynthetic plants such as mushrooms and toadstools, which derive their nutrition from other plants and animals – both living and dead; the bryophytes – mosses, liverworts and hornworts, and the pteridophytes – clubmosses, horsetails and ferns. A fifth group of common plants, the lichens, is not really a separate group at all because each 'plant' consists of a fungus and an alga living together in a symbiotic relationship – one that is to the mutual benefit of both partners. This and a variety of other relationships, both beneficial and otherwise, will be described in Chapter 17.

The algae

The algae form an astonishingly varied group of mainly aquatic plants with a very wide range of structures and life cycles. There are three major divisions, the green, brown and red algae, all of which will be familiar to a keen observer of sea-shore life.

The green algae

The green algae range in size from tiny, single-celled, free-swimming organisms to long strands of cells joined end to end, like *Spirogyra* which forms the green slime on freshwater ponds, and thin, flat sheets of tissue like the sea lettuce (*Ulva*) which has a more typical plant-like appearance. In most species there is little specialization of cells for particular functions. In *Ulva*, however, there is a slight degree of specialization because the cells at the base of the plant secrete an adhesive substance which attaches the plant to the rocks. Plants like *Ulva* have not developed specialized organs like roots, shoots and leaves, or transport systems, since virtually every cell in the flat green plant carries out its own photosynthesis and is bathed in an aqueous solution of mineral nutrients – the sea.

The green algae reproduce themselves by both the sexual and asexual processes, and they possess a bewildering range of life cycles far beyond the scope of this book to describe. In some, for example *Ulva*, the gametophyte plant looks exactly the same as the sporophyte plant, while in others, such as *Spirogyra*, the plant we see is the gametophyte, and the sporophyte generation has been reduced to just a single cell – the zygote.

The familiar *Mycena* fungus often grows in attractive clusters on fallen trees and stumps, but beneath the surface its delicate thread-like hyphae are spreading throughout the wood, causing its decay.

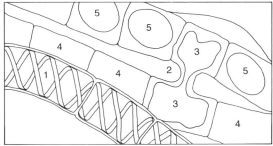

In sexual reproduction, the spiral chloroplasts of *Spirogrya* (1) break up. Projections from the walls of cells in adjacent filaments join up (2) to form linking tubes. The cell contents pass through (3) leaving the cells in one filament empty (4) while the fused contents of the others form thick-walled zygospores (5) capable of withstanding drought.

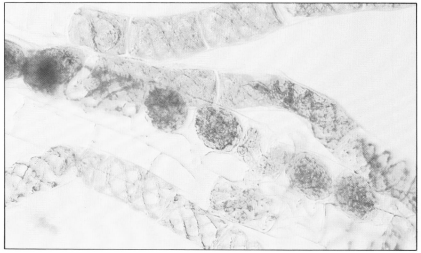

The brown algae

The brown algae are the ones that will be most familiar to the reader for they dominate the plant life of the rocky inter-tidal zone of the sea-shore. They are generally larger and more complex than the other algae: the giant kelp *Macrocystis*, for example, can have fronds more than 100 metres long, while *Sargassum*, which forms free-floating masses in the oceans, has an internal specialization of cells reminiscent of the more advanced types of land plants.

The large brown algae show distinct differentiation of their cells into specialized structures – an adhesive holdfast for attaching the plant to the rocks, a stalk or stipe, and a blade or lamina which offers a large flat area to the light. Many species, whether fixed to rocks or free-floating, have air-filled floats or bladders which keep the fronds as near to the water surface as possible so that the maximum amount of light can be intercepted for photosynthesis.

In most cases the familiar large brown plant of the sea-shore is the sporophyte generation, the gametophyte being very much smaller, in some cases reduced to only a few cells. Details of the alternation of generations differ considerably from one species to another; in *Laminaria*, for example, the leathery, strap-like sporophyte is just one of *three* forms in which the plant can exist. At other times in its life cycle it exists as quite separate and distinct microscopic male and female gametophyte plants!

The red algae

The red algae are mainly marine organisms of quite complex structure. A few are free-floating but most of them grow attached to rocks or to other algae, especially the browns. They tend to be less robust than the brown algae and for the most part live farther down the shore where they are less exposed to the air and the pounding of the waves. Some are found growing to depths of 100 metres and more. They have a thread-like or flat structure and are often intricately branched. Some produce large quantities of gelatinous material which makes them very slimy to the touch, while others accumulate calcium from the sea and deposit it within their structure as calcium carbonate. These algae are partly responsible for the formation of reefs, although many animal species participate in the process as well.

The red algae have a distinct alternation of generations in which the gametophyte and

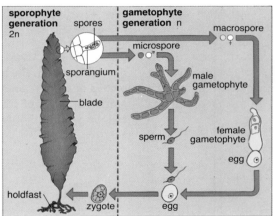

The tough rubbery fronds and branched holdfasts of *Laminaria digitata* (*above*) are a familiar sight on rocky shores.

The large sporophyte plant (*left*) has two sets of chromosomes (2n) but produces spores with only one set (n). These germinate into separate male and female gametophytes which in turn release 'one set' eggs and sperm. When these fuse they form the 'two set' zygote that will become the next sporophyte.

Most red algae grow on rocks or as epiphytes on other algae. In this case (*left*) the alga *Ceramium* has attached itself to another, more robust species called *Gigartina*.

The red alga *Corallina officinalis* accumulates calcium carbonate (lime) in its tissues and is a contributor to the rocky structure of coral reefs.

sporophyte plants are usually identical. However, a number of life cycles have recently been discovered in which plants of the two generations are quite different, and in one dramatic case turned out to involve two plants that were so dissimilar that for years they had been classified as entirely different species!

Other groups of algae

Two other groups of organisms are, or have been, included in the algae. The first group, the cyano-bacteria, used to be called the blue-green algae but it is now clear that they are more properly grouped with the bacteria. They are very important organisms because they have the capacity to 'fix' atmospheric nitrogen – a subject to which we will return in a later chapter. The second group, the chrysophyta, are a major component of the phytoplankton – the microscopic, photosynthetic, single-celled organisms that float in the upper layers of the sea and freshwater lakes. These organisms carry out a major part of the total photosynthesis that takes place on the earth's surface and so act as the primary energy-capturing stage in the food chains of vast numbers of aquatic animals.

Industrial uses of algae

The algae are an extremely useful and commercially important group of plants. Brown algae are harvested in huge quantities, especially from the great kelp beds off the California coast. Formerly, they provided a major source of iodine. Nowadays, however, they are harvested mainly for complex carbohydrate polymers such as sodium alginate and alginic acid which are used as bases and thickening agents in pharmaceutical products like toothpaste and cosmetics, in the food industry in artificial cream, and in industry to provide a base for non-drip paints and many other products. These ancient plants also serve man in the laboratory, the red alga *Gelidium* being the source of the inert gelatinous material agar, on which biologists grow their fungi, bacteria and tissue cultures.

The fungi

The fungi are often placed in the plant kingdom, but in reality they are very different from plants and should really be classified as a separate group altogether. The principal reason for this is that not one of them possesses chlorophyll, and since they cannot synthesize their own carbohydrates they obtain their supplies either from the breakdown of dead organic matter or from other living organisms. Furthermore, the walls of fungal cells are not made of cellulose like those of true plants, but of another complex sugar-like polymer called chitin – the material from which the hard outer skeletons of shrimps, spiders and insects are made. The difference between the chemical composition of the cell walls of fungi and those of higher plants is of enormous importance

The ability of fungi to break down the cellulose walls of plant cells makes them extremely damaging. Here, dry rot fungus has invaded household timbers and is destroying them.

21

The 'fruiting bodies' or spore-producing structures of fungi are wonderfully varied in form and colour. These are just some of the more common varieties. Reading clockwise from top left; *Mitrula paludosa, Amanita muscaria, Fomes annosus, Lactarius rufus, Polyporus squamosus* and the bird's nest fungus *Cyathus striatus*.

because it enables the tips of the growing fungal hyphae to secrete enzymes that break up the walls of plant cells without having any effect on those of the fungus itself. It is these cellulose-destroying enzymes that enable fungi to attack timber, cloth, paper and indeed anything made from wood, wood pulp, cotton, flax or other plant material.

If the fungi are so different from normal green plants, why then should we consider them at all in this book? They are included for two reasons. First, they enter into all kinds of relationships with plants, some undoubtedly beneficial, others quite definitely not; and second, they are an extremely important group of organisms from the economic and commercial point of view.

The destructive power of the fungi is impressive. They can attack, and live on, a huge range of materials, from wood and leather to aircraft fuel and photographic film. They are a major cause of structural damage to building timbers, a cause of disease in animals and man, and one of the greatest causes of agricultural losses. Entire crops can be wiped out by fungal attacks, both before and after harvesting. Some fungi can grow at $+50°C$ while others can grow at $-5°C$, so even food in cold storage is not completely safe from them.

But the fungi do have something to be said in their favour. They bring about the decomposition of dead organic matter, thus releasing carbon, nitrogen and other atoms for recycling. Without them the earth would soon be buried under a thick covering of dead animal and plant remains! Fungi also enter into a number of beneficial partnerships with plants (Chapter 17), and they are the source of many of the most potent antibiotics used in clinical medicine, one of the most familiar being penicillin. In addition, fungi are used to produce some of the finest types of cheese, while many species are themselves eaten for their delicate flavour. The yeasts, too, are fungi, and it is on their ability to produce alcohol (Chapter 12) that the wine, spirits and beer producing industries are based.

The fungi grow as long, thin, microscopic filaments called hyphae, which may be simple threads, or branched to form a mat called a mycelum. They can also be closely packed together to form some of the larger and more familiar reproductive structures like the mushrooms and toadstools. Fungi indulge in both sexual and asexual reproduction, but the details of these processes are far more complex than we need in this brief introduction to the major divisions of the plant kingdom.

The mosses and liverworts

The mosses, liverworts and hornworts – the bryophytes – are a moderately successful group of plants with a relatively simple internal and external structure. They are mainly confined to wet or damp places, a limitation imposed by their lack of a protective outer layer to prevent water loss, and by their possession of delicate, free-swimming male gametes, which require a film of water if they are to reach and fertilize the egg cells.

The gametophyte plant – the independent and most obvious generation – has no roots, just single or multicelled filaments that serve to anchor it to the ground. In many species there is a superficial differentiation into stem and leaf, but the leaves are not like those of higher plants and the stem does not have the normal transport systems found in higher plants. Essentially, each cell absorbs its own water and mineral requirements from its damp surroundings, and provides its own carbohydrate nutrients by photosynthesis – although some has to be supplied to the stem and rhizoid cells from the photosynthetic leaves. In many bryophytes the gametophyte plant consists of a flat green structure rather like a small seaweed.

The sporophyte generation is totally dependent on the gametophyte and consists of a foot, embedded in the gametophyte, and a stalk, at the end of which is a large hollow structure called a capsule or sporangium. This contains the spores which, when liberated, germinate to give a new gametophyte plant. Two examples will illustrate the range of structure and organization in the Bryophyta, the moss *Mnium hornum* and the liverwort *Marchantia*.

Marchantia and some other liverworts have an unusual method of asexual reproduction. Cup-like structures develop containing spherical green bodies called gemmae. These bodies, looking rather like miniature bird's eggs in a nest, eventually detach themselves and germinate directly into new gametophyte plants. Sexual reproduction in mosses and liverworts involves the production of motile sperm which swim independently to fertilize egg cells contained in a flask-shaped structure called an archegonium. The fertilized egg develops into the sporophyte plant *in situ*, which is why the

Two generations in one. In this photograph of the moss *Mnium hornum*, the leafy green lower part is the gametophyte plant. The slender brown stalks and the green capsules containing the spores are two sporophyte plants. They have no independent existence and rely solely on the gametophyte plant for water and nourishment.

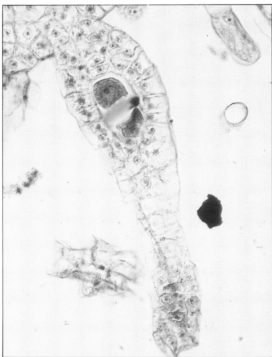

The liverwort *Marchantia* can reproduce asexually by forming gemmae (*top left*) which eventually develop into separate plants.

Other photographs show the structures of sexual reproduction, starting (*top right*) with a forest of male antheridiophores and female archegoniophores.

The archegoniophore (*above*) bears the female sex organs on its lower side, protected by the bell-shaped sheath. Closer still (*centre*) and a red-stained egg cell becomes visible inside the flask-shaped archegonium.

The antheridiophore (*centre right*) bears the male sex organs in its flattened upper surface, and (*bottom right*) high magnification reveals the sperm cells inside.

Finally (*bottom left*) after fertilization a small sporophyte plant develops under the archegoniophore.

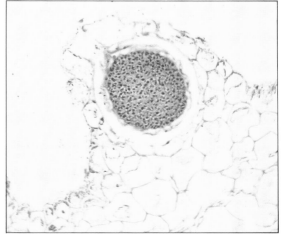

foot of the sporophyte plant is always embedded in the gametophyte, from which it derives its nutrients.

The clubmosses, horsetails and ferns

The clubmosses, horsetails and ferns, together with their fossil ancestors, seem to have solved the problems of living on land rather better than the mosses and liverworts, since they grow considerably larger and are more tolerant of drier environments. They are called the pteridophytes. Their spores have protective coats which enable them to be dispersed over a wide area, and the sporophyte, which is the dominant generation, has a complex structure. Like the more advanced plants, it has a waxy layer covering its outer surface. This layer, the cuticle, is impervious to water and so greatly enhances the plant's capacity to escape from the rather humid conditions to which most bryophytes are confined. A further advance is the presence of pores in the epidermis. These allow adequate inflow of carbon dioxide for photosynthesis, but as they can open and close, they also enable the plant to limit the amount of water loss under adverse conditions. The pteridophytes have not, however, completely solved the problem of living on land because the gametophyte plant, which is always completely independent, is small, delicate and vulnerable to desiccation. In addition, the gametes involved in sexual reproduction are largely unprotected, and the free-swimming male gamete requires a film of water in which to swim to reach the egg cell.

Another important advance found in the pteridophytes is the development of specialized transport tissues. As the sporophyte plants became larger and more complex, they needed to transport water from their roots to their leaves, and the sugars made in photosynthesis from the leaves to the non-photosynthetic parts of the plant. These transport tissues, the xylem and phloem, are described in Chapter 10. Roots, too, are found in all but the most primitive pteridophytes.

While sexual reproduction is limited to the gametophyte, vegetative reproduction can occur in several ways in the sporophyte. Parts of the branching stem, which often grow horizontally under the soil, may become separated, either by being broken or by the older parts dying, and thus begin an independent existence. Some ferns have developed the capacity to grow young plantlets on the margins of their

The fern *Asplenium viviparum* can reproduce asexually by developing miniature plantlets on the edges of its fronds. Eventually these drop off and develop into independent new plants.

fronds (leaves), and these simply drop off at maturity and become independent plants.

The pteridophytes dominated the landscape 300 million years ago and it is principally their fossilized remains that form the coal and oil deposits of today. They were ideally suited to the primaeval swamps and other damp habitats of 300 million years ago and their decline was largely due to climatic changes which led to the demise of such habitats. However, they are still a very successful group, the bracken fern being one of the few plants said to grow in every country in the world!

The world's first land plants

The most primitive pteridophytes known are those from the fossil record, and they probably reflect the type of plant that first colonized the land. A small plant called *Rhynia*, found in fossil beds in Aberdeenshire, Scotland, dates from about 400 million years ago. It consists of a stem whose only branching pattern is a division into two equal shoots, and this so-called dichotomous branching pattern is found in all primitive pteridophytes.

Rhynia had neither leaves nor roots, but its aerial stems were attached to a horizontal stem, or rhizome, and were covered with a waxy cuticle. The stems also had stomata, and presumably functioned as photosynthetic organs. Internally *Rhynia* stems had a central zone which contained a strand of xylem, and cells that looked like phloem. Other fossil plants from this period appear to show the development of leaf-like structures on a stem that was otherwise similar to that of *Rhynia*.

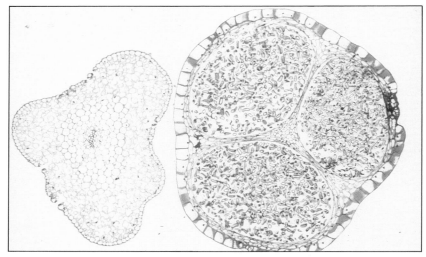

The most primitive and simple present-day pteridophytes are the two species *Psilotum* and *Tmisipteris*, both of which can be regarded as living fossils. *Psilotum* is a terrestrial plant which grows in tropical regions. It is leafless and rootless and consists of a horizontal rhizome and a dichotomously branched, green, aerial stem having bright yellow sporangia, fused together in threes, located on very short branches at intervals along its length. The other most primitive living pteridophyte, the small fern-like *Tmisipteris*, grows on the trunks of tree-ferns in Australia and New Zealand. It has a pendulous leaf-like structure which seems to be the principal site of photosynthesis, and its sporangia are fused together in twos.

The clubmosses and horsetails
Other present-day remnants of plant groups that had their heyday in the geological past are the lycopods (clubmosses) and the horsetails. The great primaeval forests that gave rise to our coal and oil fields consisted mainly of the huge tree-like *Lepidodendron* and *Sigillaria*, which are related to the present-day clubmosses *Lycopodium* and *Selaginella*, and *Calamites* which is related to the present-day horsetail *Equisetum*.

Fossilized stems of the giant clubmoss *Lepidodendron* show that they were covered with leaves, because the scars left when they died are clearly visible. In addition, they had reproductive structures called cones, found in most present-day clubmosses, in which the sporangia were concentrated and protected. Fossil stems of *Calamites* show a very intricate pattern of stem ridges, and had their leaves arranged in distinct and separate rings, or whorls, as found

Even though *Psilotum* (*above left*) is a very primitive plant, a cross section through the stem and one of the sporangia (*above*) shows that it does have specialized fluid-carrying tissues (xylem and phloem). The section also shows that the sporangia are fused together in threes.

A glimpse into the past. Based on fossil evidence, this is what the great primaeval forests of the Carboniferous Period must have looked like. The two dominant species are the tall branched *Lepidodendron* and the shorter, conical, *Calamites* horsetails.

today in the only survivors of this group – the horsetails. The root system of these large pteridophyte plants was not very extensive, and because of the need for a wet environment for the gametophyte to complete sexual reproduction, they disappeared very quickly once the swamps dried up between 250 and 200 million years ago.

The present-day lycopods, such as *Lycopodium*, are small plants with a rhizome, roots, and aerial branches densely clothed in small

The surface of a fossilized *Lepidodendron* stem (*far left*) clearly shows the scars where the leaves were attached. Even the places where the vascular bundles passed from the stem into the leaf stalk can be seen in the centre of the diamond-shaped leaf scars.

This superbly preserved cone (*left*) is believed to be from a *Lepidodendron* tree, but as not one specimen has ever been found actually attached to a *Lepidodendron* branch, the cone is given its own name—*Lepidostrobus*!

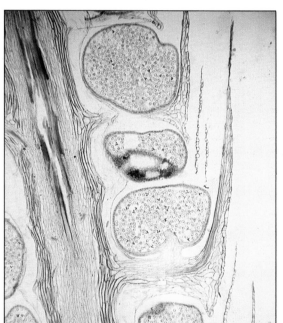

The clubmoss *Lycopodium clavatum* (*far left*) bears its yellowish cones on short stems. If the cone is split lengthways (*left*) and viewed under a high powered microscope, the sporangia borne on the upper surface of each sporophyll, can be seen to contain just one kind of spore. The sporophyll is a small leaf specially modified to support and protect the sporangia.

leaves. In *Lycopodium* we can perhaps see the first attempts by plants to concentrate their reproductive apparatus into a specialized structure, in this case, a cone. In the *Lycopodium* sporophyte, the sporangia develop on the upper side of the leaf, that is the side next to the stem. Leaves bearing the sporangia are called sporophylls. In *Lycopodium selago*, zones of sporangia-bearing leaves alternate with zones of sterile leaves along the stem. In *Lycopodium clavatum*, however, the sporophylls are con-fined to the region at the tip of a stem, and since the part of the stem immediately below is elongated, the sporophylls form cone-like structures on the end of stalks. The sporophylls are slightly modified in shape compared with normal leaves, but when the cone-like structure is viewed in longitudinal cross-section the similarity to the cones of pine trees and cycads is obvious. Living *Lycopodium* species produce only one type of spore, and the gametophyte plant bears both male and female sexual organs.

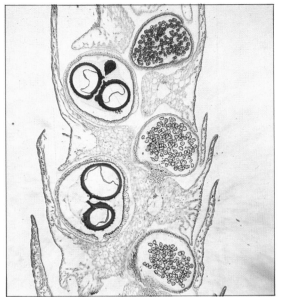

Selaginella is a small herbaceous plant which often has a curious pattern of leaf arrangement in which one leaf of each pair on the stem is larger than the other. In the next pair along, the relative sizes are reversed. *Selaginella* also has sporangium-bearing leaves, or sporophylls, concentrated into cones, but unlike *Lycopodium* it produces two kinds of sporangia within a single cone. One contains only microspores, which divide to produce the male gametophytes. The other produces megaspores, which divide to produce the female gametophyte, which is retained within the spore case. When the microspores land on or near the female gametophyte they burst and liberate their sperm, which then fertilize the egg cells. The fertilization process can occur after both the microspore and macrospore have been shed from the cone, but sometimes the megaspore germinates and produces egg cells while still in the cone, where they can be fertilized by the contents of an incoming microspore. In some species the fertilized egg can immediately start to develop into a new sporophyte plant, and since all this can occur *inside* the original cone, this reproductive process is superficially very similar to that found in the present-day flowering plants (Chapter 4). It seems, then, that in the course of evolution, plants probably experimented more than once with the production of two types of spore. The lycopod experiment, represented today by *Selaginella* and a few aquatic ferns, is just one of those experiments – one that proved partly, though not wholly, successful.

Despite being one of the dominant plant groups 300 million years ago, the horsetails are today represented by only a single genus, *Equisetum*. The sporophyte is a large, complex and beautiful plant which grows best in lush, damp situations. Its ancestor *Calamites* was a tree which commonly grew to 30 centimetres in diameter and more than 20 metres in height.

A shoot of the modern horsetail *Equisteum* (*left*) shows the characteristic arrangement of branches in rings or whorls along the stem. Deep scars encircling the fossilized stem of a *Calamites* plant (*below*) show that this primitive relative shared the same distinctive feature.

A close-up of *Equisetum* shows the green branches arranged in a ring, the small brown scale leaves pressed against the stem, and the prominent parallel stem ridges. Compare these features with the fossil *Calamites* stem opposite.

unusual, being largely hollow with a number of secondary canals, presumably to allow gases to diffuse through the tissues.

The reproductive structures of the sporophyte are well developed cones which occur at the end of the green stems in some species, or on specialized white shoots, devoid of chlorophyll, in others. The cones are very interesting structures in which the sporangia are afforded a high measure of protection until they are ripe. Each cone consists of a central stem on which are arranged rings of mushroom-shaped projections called sporangiophores, and the sporangia project inwards towards the centre of the cone from the underside of these protective mushroom-like structures.

All the horsetails, fossil and living, have parallel ridges along their stems, and their branches and leaves are arranged in rings or whorls. The leaves are brownish-black scale-like structures that are pressed tightly against the stem. By contrast, the main stem and its branches are green, and they alone carry out photosynthesis. The internal structure of the stem is also very

The ferns

The ferns are a very widespread group of plants whose sporophytes, the dominant plants, show a remarkable range of structure. Some, like the aquatic ferns *Marsilea*, *Pilularia*, *Salvinia* and *Azolla*, would scarcely be recognized as ferns at all. *Marsilea* looks like a four-leafed clover, *Pilularia* appears simply as leafless green stalks, while *Azolla* and *Salvinia* are free-floating plants that live on the surface of lakes. The complexity of the structure that can be found

The cone of *Equisetum arvense* (*right*) is borne at the end of a special short shoot which has no chlorophyll but does have rings of brown scale leaves. The sporangiophores are arranged in rings around the axis of the cone, and if the cone is cut in half (*far right*) their intricate structure can be seen. The sporangia are carried beneath the mushroom-like caps of the sporangiophores and so project inwards, towards the centre of the cone.

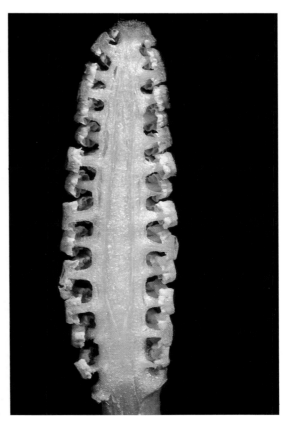

The floating water fern *Salvinia* (*above* and *right*) has two green surface leaves with spiky upper surfaces and one finely branched submerged leaf, occurring in groups at intervals along a slender horizontal stem.

One of the most beautiful and intriguing of plant shapes is the uncurling frond of a fern. Here, in spring, fronds of a young male *Dryopteris filix-mas* fern display the crozier-like form that protects the young leaflets.

is illustrated by *Salvinia*, which appears to have two leaves and a finely branched root system arising at intervals along the length of its horizontal stem. In fact, there are three leaves at each joint, but one is finely divided into thin, branched filaments and hangs down into the water where it looks like, and presumably functions as, a root system – true roots being absent in this species.

The land ferns can vary in size from the small species found on walls, such as *Asplenium ruta-muraria*, and the damp-loving filmy fern *Hymenophyllum*, to the huge tree-ferns *Cyathea* and *Dicksonia* found in Australia, New Zealand and South America. The stems may be upright or horizontal, and have both roots and leaves, which in the ferns are called fronds.

The reproductive structures of the sporophyte may be borne on specialized, modified fronds called fertile spikes, or more commonly on the underside of the normal green photosynthetic frond. These sporangia are arranged in groups called sori and are usually, but not always, protected by a structure called an indusium. Whether or not the indusium is present, its shape, and the position and arrangement of the sori on the lower surface of the fronds, are the principal features used in recognizing different species of fern. Each sporangium is borne on a little stalk and many have what amounts to an explosive device to ensure that when it ruptures it distributes the spores as far as possible from the mother plant.

The land ferns all produce a single, fragile gametophyte plant which bears both male and female sex organs. In many of the common ferns, such as *Dryopteris*, the gametophyte is heart-shaped. It is rarely larger than a finger-nail, and bears sex organs on its lower surface, in addition to thread-like rhizoids which attach it to the soil. The multiflagellate sperm swim to effect fertilization, and the young sporophyte which develops from the fertilized egg may for a while be nourished by the photosynthetic capacity of the 'mother' gametophyte, although it soon develops its own roots and fronds and becomes independent.

The pteridophytes are all vascular plants, that is, their sporophyte plants have strands of vascular tissue which contain specialized xylem and phloem (Chapter 10). In this respect they differ from the algae and bryophytes. On the other hand, they differ from the more advanced gymnosperms and angiosperms in that they do not form seeds, but rather produce spores which give rise to a wholly independent gametophyte generation.

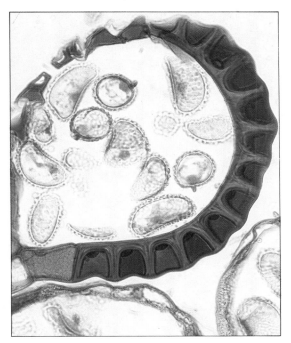

Groups of fern sporangia occur in clusters called sori on the undersides of the mature leaf fronds. *Dryopteris* sori (*top left*) are protected by a kidney-shaped grey indusium, those of the maidenhair fern (*Adiantum capillus-veneris, above*) are partly protected by the curling edges of the leaf-fronds, while those of the common polypody (*Polypodium vulgare*) have no indusium so the yellow sporangia can be clearly seen.

The detail (*left*) of a *Polypodium* sporangium shows the thick walls of the annulus (red) which, when ripe, will catapult the spores into the air.

CHAPTER 4

Variety in Life-styles: The Seed Plants

The two groups of seed-producing plants are the gymnosperms – the conifers of temperate regions and the cycads of the tropics, and the angiosperms – the flowering plants, including all the grasses and cereals. This group contains virtually all the crop plants grown for food.

The success of the seed plants over their seedless relatives is due to three major advances in their organization and life-style. These advances amount to quantum leaps in evolutionary progress and made possible the colonization of many new environments. First, the seed plants have dispensed altogether with the vulnerable, free-swimming male gamete which confines the seedless plants to environments that are at least periodically wet and humid. Instead, the male gametes are enclosed within protective coats to form pollen grains. These are sufficiently robust to be transported long distances under dry conditions, and so are able to fertilize egg cells on plants many kilometres away. Second, they have devised elaborate structures to protect their female gametophytes and egg cells. Third, they have developed a system for releasing their offspring in neat protective capsules called seeds – each seed consisting of a tough outer case enclosing a young sporophyte plant, or embryo, and a supply of food to give it a start in life. A further advance is that the seed-borne embryo almost always has a dormant phase during which it will not germinate even in the most favourable conditions (Chapter 5). It is this delay in development and the protective nature of the seed case that have enabled seed plants to disperse themselves so successfully over enormous distances.

These two major plant groups, the gymnosperms and the angiosperms, take their names from the way in which their female reproductive structures develop. The word gymnosperm means 'naked seed', while angiosperm means 'enclosed seed'. In the gymnosperms, the egg cells develop on scales held on the woody plates of the familiar cone, but in angiosperms they are always completely enclosed within ovaries which, after fertilization, develop into fruits, many of which are of commercial importance.

The gymnosperms

The gymnosperms are easily recognizable because with a few exceptions such as the yews,

Cycad cones are typically larger and much softer in structure than those of coniferous trees. This huge male cone, growing on a *Macrozamia* tree, is 40cm long and 20cm wide.

The green seeds of the yew (*Taxus baccata*) are surrounded when ripe by vivid red fleshy arils, which make them very attractive to birds. The yew is one of the very few gymnosperms that does not form female cones.

the male and female reproductive structures are cones, and in many species these are large, conspicuous and very beautiful. The gymnosperms are divided into three groups: the conifers, which are almost all trees of temperate regions, mostly with needle-like or rather small leaves; the cycads, which are somewhat smaller tropical plants with elaborate palm-like leaves and exceptionally large cones; and lastly, a rag-bag group consisting of four very odd plants which have some of the features of the gymnosperms and some belonging to the angiosperms. It is a group that has clearly evolved, or was created, especially to defeat a neat classification of the plant kingdom! This group includes the *Ginkgo*, a beautiful tree with fan-shaped leaves, often grown in ornamental gardens, and *Welwitschia*, which appears to achieve the impossible by growing only in the hottest and driest place on earth – the otherwise barren Namib desert of South-West Africa.

The gymnosperms include the largest living organisms the world has ever seen, and also the world's longest-living organisms. Individual specimens of *Welwitschia*, and of the giant coast redwood trees of California and Oregon, are thought to be over 2,000 years old; bristlecone pines in the Arizona desert have been dated at 4,000 years old, while one cycad specimen in Australia is estimated to have already celebrated its 5,000th birthday!

The conifers

This economically important plant group provides more than 75 per cent of the timber used in the building industry. It is also the major source of pulp for paper, and the resins produced by the trees are important raw materials in the chemical industry. Most of the conifers are tall trees, with woody stems and small or needle-shaped leaves that are specially adapted to withstand dry conditions. Most are also evergreen, which means that they do not shed all their leaves simultaneously at one particular time of year. In fact, they shed leaves continuously, but since others are developing at the same time, they are always well clothed in leaves and can carry on photosynthesis throughout the year.

The success of the conifers is due in part to their tall woody stems, which enable them to grow above other plants and so compete successfully for the available sunlight. The strong trunk of the tree is formed by a process of secondary growth which produces a huge

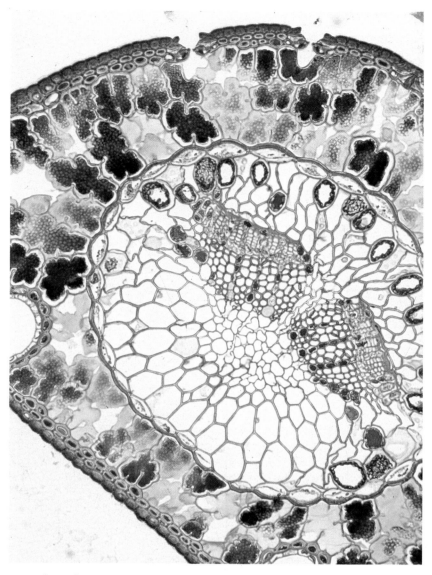

quantity of wood – the remains of what were originally water-conducting cells. This process of secondary thickening is described in Chapter 10. The leaves of the conifers are typically narrow and elongated and have a special structure that enables them to function without damage through the changing seasons of the year. They have a low surface-to-volume ratio, which reduces the surface area from which water can evaporate in dry, windy weather. They are also covered with a thick waxy cuticle, and the pores, or stomata, in the surface layer of cells are sunk into depressions, which further reduces evaporation.

Male and female cones develop quite separately on the tree, the male cones on the lower branches and the female on the upper ones. The female cone is the larger and consists of a

A section through a pine needle (*above*) shows the central vascular system (here stained red and light blue) through which fluids are transported. Photosynthesis occurs in the outer ring of light blue cells. Also visible are large circular resin canals and the leaf pores (stomata) sunk into depressions in the surface.

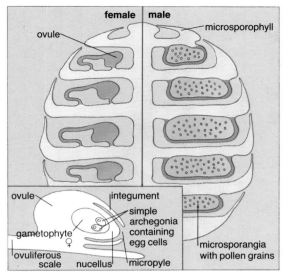

female | male

ovule
microsporophyll

ovule
integument
simple archegonia containing egg cells
gametophyte ♀
microsporangia with pollen grains
ovuliferous scale
nucellus
micropyle

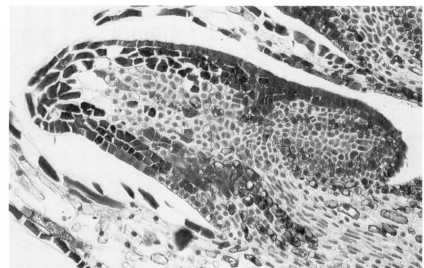

central axis having arranged around it numerous sporophylls – leaf-like structures that are specially modified to bear sporangia. Each sporophyll bears a thin scale on its upper surface, and it is at the base of this scale, well protected deep within the cone, that two ovules develop. A fascinating sequence of changes then takes place, by the end of which each ovule consists of a cluster of fertile egg cells enclosed in a mass of tissue called the nucellus, all neatly wrapped in a protective sheath. Later, this sheath will form the tough seed case, but at this stage it is still soft, and pierced at one point by a narrow channel called the micropyle – the access point through which, eventually, pollen grains will make their way to fertilize the egg cells. The principal features of male and female cones are shown above.

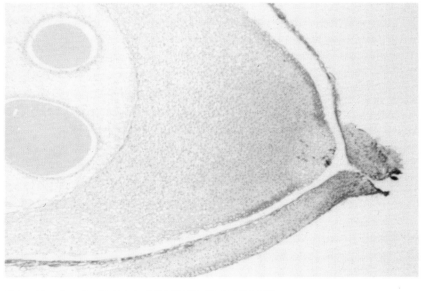

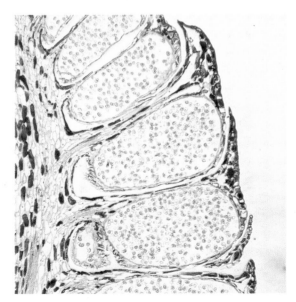

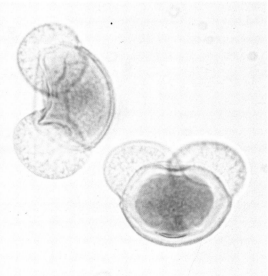

Reproductive structures of the pine trees. Ovules (*top*) form on the upper surfaces of female cone scales (*see diagram*). The female gametophyte grows from one of the megaspores in the nucellus, and develops simple archegonia, two of which are visible in the photograph above. Also clearly visible are the enclosing integuments and the micropyle pore. Pollen forms in sporangia that develop on the underside of male cone scales (*far left*), and dispersal is assisted by small balloon-like wings (*left*).

The male cone is smaller than the female but is similar in that it also has a central axis with outwardly projecting sporophylls. Each sporophyll develops on its lower side two microsporangia which at an early stage contain thousands of microspore mother cells. These then divide repeatedly by the process of meiosis to give vast numbers of microspores, each of which develops into a pollen grain. But it is what is going on *inside* the pollen grain that is important from the point of view of reproductive evolution. When it is first formed, the pollen grain comprises a single cell. This is the first cell of the male gametophyte, and it undergoes division within the pollen grain to produce four cells – two prothallial cells, a generative cell and a tube cell. It is at this stage that the pollen grain is released and dispersed in the wind.

Pollination takes place when the wind-borne pollen gets into the female cone and is caught on the sticky fluid around the micropyle of the ovule. Some months elapse before further development takes place, and during this time the sticky fluid evaporates and the pollen grains are drawn in through the micropyle until they lie on the surface of the nucellus. Here they germinate, and each one produces a pollen tube which grows down into the nucellus, tunnelling its way towards the egg cells. Once the pollen tube begins to develop, the generative cell divides to produce a stalk cell and a body cell, which together move down to the growing tip of the tube. The body cell nucleus then divides to produce two naked sperm nuclei, and when the pollen tube has grown through the nucellus and into the neck of the archigonium, the tip ruptures and the male nuclei are released to fuse with the eggs. There is usually a long interval, often of many months, between pollination, the arrival of the pollen at the micropyle, and fertilization – the actual fusion of the male and female gametes.

Once fertilization has taken place, the seed begins to develop. The process of cell division is rather involved, but the end result is a recognizable little conifer seedling, with a ring of embryonic needle-like leaves, a shoot and a root, all embedded in a mass of tissue and surrounded by the seed coat. When ripe, the seed coat is thin, and remains attached to the scale on which it started life. The scale itself becomes thin and papery, forming a large wing which aids dispersal of the seed by the wind when it is finally released from the cone.

The cycads

The cycads are a primitive and ancient group of gymnosperms, somewhat like the palms in general outline but quite different from them because the palms are flowering plants. The upright stems of cycads usually have a crown of very large leaves, often edged with unpleasantly sharp spikes. Details of their reproductive structure and mechanisms are very like those of the conifers, but they retain one characteristically primitive feature. In the cycads, and in *Ginkgo*, the two male gametes formed in the pollen grain are not naked nuclei as in the conifers, but instead are motile, swimming sperm like those found in the more primitive plant groups described earlier. These gametes are, however, never really free because they can swim only within the confines of the pollen grain and pollen tube, from which they escape only when the tip of the tube ruptures on reaching the egg cell.

The misfits

In the remaining group, the rag-bag collection of misfits, there is so much of functional and behavioural interest. The beautiful *Ginkgo* tree is rather like the giant panda – more often seen

In *Cycas revoluta* the large green ovules are carried on the edges of leaves that are only very slightly modified. The plant does not form anything remotely like a female cone, but it does illustrate the point that the sporophylls of cones in the gymnosperms are really just highly specialized leaves, modified to support and protect the structures that produce the plant's eggs and pollen.

in captivity than in the wild. Only the male tree is grown, however, because the female produces seeds that emit a nauseous smell that is well beyond human tolerance. The ovules in *Ginkgo* are produced not in cones but on the ends of short stalks which occur in pairs along the stem – a situation reminiscent of the arrangement in the yew and perhaps indicative of that tree's very primitive nature.

Gnetum is an interesting plant in that it has broad leaves with a vein pattern similar to that found in the broad-leafed flowering plants. Its cone-like reproductive structures are also rather like flowers in that both male and female sporangia may develop in the angle between a sporophyll and the cone axis. However, it is extremely rarely that both mature – usually it is one or the other. In complete contrast, *Ephedra* has only scale leaves like the horsetails, all photosynthesis being carried on by the jointed stem. The male and female cones of this species are borne on separate plants.

Welwitschia is a bizarre plant that consists of a very deep-rooted combined stem and root, shaped like a very long inverted cone or large woody parsnip. The top of the stem is concave, woody, and disc-shaped. Despite living for

Two of the most unusual gymnosperms are *Ginkgo biloba* (*above left*), a tree from China with very beautiful fan-shaped leaves, and *Welwitschia* (*above*) which lives in the Namib Desert and can survive for thousands of years despite never having more than two leaves!

This close-up (*left*) of the top of a *Welwitschia* plant reveals an unusually large number of male 'flowers' at the top of the stem. The 'flowers' are really the male cones.

2,000 years, the plant produces only two leaves, which spread across the hot sand of the desert, the distal ends usually turning yellow and dying, giving the plant a very untidy appearance. The secret of the longevity of these leaves, however, is that they grow continuously from their bases, producing more and more leaf as the years go by. The only other plants in which this pattern of leaf growth is found are the monocotyledons, particularly the grasses and cereals. *Welwitschia* grows in an area where it may rain only once in five years or so, and the wonder is that it survives at all, despite its very deep root.

These last three plants are clearly gymnosperms because they have naked ovules, but taken together they show many features found in the flowering plants. Whether they really represent a linking group between the gymnosperms and angiosperms is a matter for speculation: the concensus at the moment is that they are at the dead end of one or more unsuccessful lines of gymnosperm evolution.

The angiosperms

All the plants in this group develop flowers, which may be large, beautiful and brightly coloured like those of the orchids and tulips, or small and inconspicuous like those of the grasses and many small, floating, aquatic plants. A flower is a very precise structure, best thought of as a central stem bearing a number of leaves which are highly modified to facilitate sexual reproduction. The essential feature of the flower, and the one that distinguishes it from the reproductive structures of other groups of plants, is that the ovule or ovules are totally enclosed in, and protected by, a structure called an ovary or carpel.

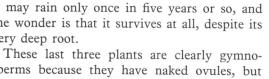

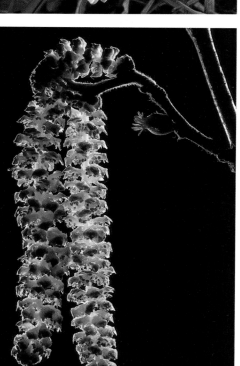

There is simply no limit to the variety of shape and colour displayed by flowers. Illustrated here, clockwise from top left, are the stunning flower of *Epiphyllum*; the compound flower of a daisy, consisting of scores of minute florets; the male (catkin) and female (pink) flowers of the hazel (*Corylus*), and a strikingly beautiful lily with petals curled back to reveal stamens with rust-brown anthers at their tips and a long green style ending in a sticky white stigma.

The angiosperms are a tremendously successful and widespread group of plants. They dominate this planet with over 250,000 species
described so far and possibly half as many again
yet to be discovered. They show astonishing
variation in structure, and exploit virtually
every kind of habitat. Some are enormous, like
the flowering trees of acacia, eucalyptus and
oak; others are minute, floating plants, like
the duckweeds *Lemna* and *Wolffia*. Some are
herbaceous and live for only one year; others
are woody and live for hundreds. Some have
very flexible stems and have adopted a climbing
habit, developing specialized structures with
which they attach themselves to other plants
for support. Some have no chlorophyll, and
have to live as parasites on normal plants,
while others catch insects and other animals to
provide the nitrogen that is totally lacking in
the soil in which they grow. Some flowering
plants are specially adapted to live in deserts,
while others live wholly submerged in water.
Each such specialization represents the solution
of a specific problem, some of which have been
solved by beautiful structural modifications
while others have necessitated changes in biochemical and physiological processes, some of
which will be discussed in the course of this
book.

There are two major divisions of the flowering plants; the monocotyledons and the
dicotyledons. Cotyledons are the first leaves on
the stem of an embryonic plant, and as the
name implies, the embryo of a monocotyledon
has just one leaf while that of a dicotyledon
has two. These highly specialized first leaves
may not look like leaves at all because they are
often swollen into bulbous structures whose
function is simply to act as a food store. They
nourish the embryo from the time the seed
germinates to the time when the young plant
emerges from the ground and holds up its first
food-producing leaves to the sun.

The monocotyledons form a huge group of
plants that includes all the grasses, cereals,
palms, lilies, orchids, tulips, crocus, and so
on – the first two producing most of the food
required to sustain human life. It is rather
difficult to examine the seeds of a plant to
determine whether it is a monocotyledon –
some are extremely small and some plants
produce them only rarely. There are, however,
other characteristics that indicate whether or
not a plant is a monocotyledon. First, the leaves
are usually long and narrow, and the veins

are parallel, without side branches or net-like
arrangements; second, the arrangement of their
stem tissues is characteristic in that there are
many vascular bundles (see Chapter 10) scattered evenly throughout the stem; and third,
they do not posssess the capacity for normal
secondary stem growth – that is, unlike the
conifers and the other group of angiosperms,
the diocotyledons, they do not produce wood.
Some monocotyledons such as *Yucca* and the
palms do grow thick stems, but these develop
in a quite different way and have no solid core
of wood.

Prolific growth of the floating
water hyacinth (*above*) is a
common cause of blocked
water supplies and drainage
systems in tropical countries.

The hydrangea's broad leaf
and branching vein pattern
(*below left*) are typical of dicotyledons, whilst the narrow
leaf and parallel veins of a
grass are characteristic of the
monocotyledons.

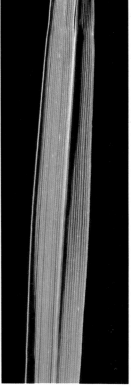

The cotyledons of dicotyledonous plants are usually extremely swollen and contain all the food reserves required for the early growth of the embryo into a young seedling. If you are in any doubt about what a cotyledon is, examine a pea. It splits into two hemispheres if the seed coat is removed, and these two hemispheres are the two cotyledons. If you look carefully you will see attached to one of them (originally to both before you tore it apart) a minute stem and root axis. These cotyledons are therefore leaves modified for food storage – they will never develop into normal foliage leaves or carry out any significant amount of photosynthesis, but will simply wither away as their contents are used up. Cotyledons are what we eat when we feed on peas and beans. Other important dicotyledonous food crops include potatoes, beets, most of the fruits we eat, and our main beverages, tea, coffee and cocoa.

Again, other characteristics of a plant will indicate whether or not it is a dicotyledon. The leaves are usually broad, and the veins in the leaves are branched or even net-like. Internally, the young stem has a single ring of vascular bundles, and in many dicotyledons the stem later undergoes secondary thickening, by which it is stiffened by the development of a more or less solid woody core. This is taken to the extreme in long-lived trees, but it also happens in shrubs and to a limited extent in herbaceous dicotyledons.

The flower

All angiosperms have a flower of some sort, and since the plants depend on their flowers for sexual reproduction, we should take a closer look at this characteristic structure and how it works.

A flower is essentially a series of modified leaves arranged in four rings around a short stem. Starting from the base, those of the first ring are usually, but not always, rather leaf-like in being green. They are called sepals. The leaves of the next ring, usually highly modified in shape and colour, are called petals. If the sepals and petals are identical, highly coloured and petal-like, as they are in a tulip or crocus flower, the whole lot is called the perianth. Neither the sepals nor the petals are involved in the production of gametes, but rather have a cosmetic function in making the flower attractive to insects and other pollinating animals by their colour, and by the production of perfume and a sugary solution called nectar.

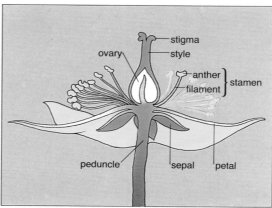

The bisected *Hypericum* flower above shows the four concentric rings of specialized leaves that make up the sepals, petals, stamens and ovaries of a typical flower.

Immediately above the petals lies the first ring of leaves modified to produce reproductive cells, in this case pollen grains. These leaves consist of a long stalk with four microsporangia at the end. The whole structure is called the stamen, and it consists of a stalk, the filament, and the anther. The four microsporangia are contained within the anther which, when mature, splits to release the pollen. The final, uppermost, ring of modified leaves forms the ovaries, of which there may be one or many, either separate or fused together. Within each ovary there develops one or more ovules. The end of the ovary develops a stalk, the style, which has a swollen end which is often coloured and always sticky. This structure is called the stigma, and it is on the stigma that pollen must land in order to effect pollination.

What sometimes makes a particular flower difficult to understand in terms of this basic four-ring structure is that firstly, parts may be missing; secondly, parts may be fused together;

and thirdly, the portion of the stem on which the modified leaves are arranged, the receptacle, may be very foreshortened, flat or even cup-shaped, so that the ovary may appear to be lower down the stem than the other parts of the flower. Endless variations on this theme can be found, but if the basic structure is kept in mind it should be possible to work out the arrangement in any particular flower.

Gamete formation, pollination and fertilization
Each anther contains four cavities or microsporangia, and within each of these a large number of microspore mother cells are produced. Each microspore mother cell then divides to produce four microspores – the pollen grains that will contribute one-half of the chromosomes needed for the production of a new generation. The wall of a pollen grain is a complex and very important structure. The inner wall is made up of carbohydrate polymers including cellulose, but the outer wall is made of very resistant substances in which there are large amounts of highly specific protein material. These proteins are very important because they provide a recognition signal by which a stigma can recognize at once if the pollen received is of its own species, and indeed whether it is from the same individual plant. This is critically important in preventing self-fertilization, and also in preventing fertilization of the egg cells by pollen of the wrong species, which would result in the formation of hybrids which are usually sterile. It is the protein in the pollen wall that sets off the unpleasant reactions that hay-fever sufferers know only too well.

When a pollen grain is released from the microsporangium it contains two cells, the original single cell having divided. One cell is called the tube cell and the other the generative cell. The male gametophyte in angiosperms is thus reduced to a two-celled structure, totally enclosed within the pollen grain. The generative cell then divides to provide the two male gametes, one of which eventually fuses with the egg cell to effect fertilization, while the other does something else which is very important and characteristic of the angiosperms, and will be described later.

The ovule in an angiosperm arises as a bulge of tissue, the nucellus, on the inside of the ovary wall and develops into a somewhat complicated structure. The bulge of tissue develops, near its base, two enveloping flaps of tissue called integuments (the gymnosperms, you may

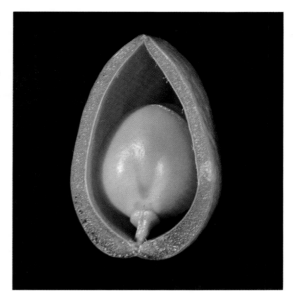

The pea in its pod (*left*) is really a seed enclosed in a fruit. Before fertilization, the seed was an ovule and the fruit (pod) was the ovary. Note the short stalk through which nutrients pass from the mother plant into the seed for storage in the cotyledons, and the tiny embryonic root visible through the transparent seed coat.

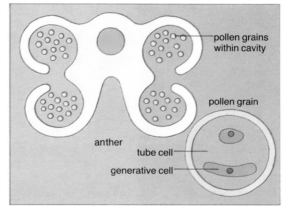

A cross-section through an individual anther shows the four pollen sacs just as their walls have split to release the pollen. Each grain (*inset*) contains a tube cell and a generative cell, enclosed in a tough outer case.

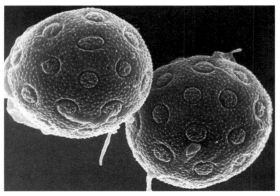

The beautiful and highly individual sculpturing of a pollen grain's outer wall is a unique feature of each plant species. These grains, reading anticlockwise, are from ragged robin (*Lychnis floscuculi*), yarrow (*Achillea millefolium*) and willow (*Salix* sp.).

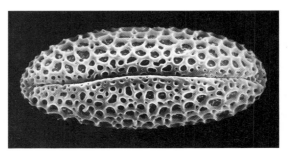

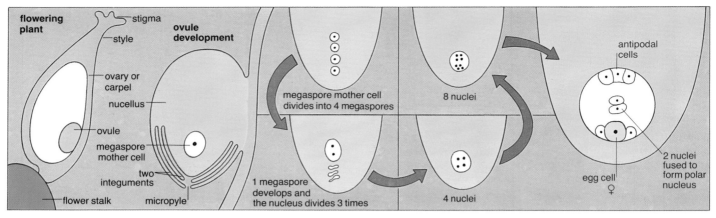

flowering plant
— stigma
— style
ovary or carpel
nucellus
— ovule
megaspore mother cell
two integuments
— flower stalk
micropyle

ovule development

megaspore mother cell divides into 4 megaspores

1 megaspore develops and the nucleus divides 3 times

8 nuclei

4 nuclei

antipodal cells

egg cell ♀

2 nuclei fused to form polar nucleus

The first two drawings above show the location and internal structure of the ovule. The remaining drawings form a sequence, starting at top left, to show the division of the megaspore mother cell into four megaspores, and the further development and division of one megaspore to give rise to a number of nuclei, the most important being the egg and the polar nucleus.

The diagram below focuses on what happens when the male gametes are released. One fertilizes the egg while the other fuses with the polar nucleus to form a special nucleus with three sets of chromosomes.

recall, had only one) and these grow up over the nucellus, completely covering it except for the minute pore at the end called the micropyle.

A large cell then develops in the nucellus, and a complex sequence of cell and nuclear divisions and rearrangements, eventually results in the production of a mature ovule, the principal features of which are the egg cell and the polar nucleus. The polar nucleus is formed by the fusion of two nuclei and thus has two sets of chromosomes, whereas the egg cell has but one.

Pollination is brought about by a pollen grain arriving on the stigma where the sugary solution promotes its further development. A passage has to be developed to channel the male gametes, which are essentially naked nuclei, to the egg cell, which may be as much as five centimetres away from the stigma in some large flowers. The pollen grain develops a long tube

– the pollen tube – which grows down into the tissues of the stigma and along the length of the style. The tube eventually penetrates the ovule by passing through the micropyle and then through the nucellus to the egg cell itself. The tube nucleus and the male gametes pass down the pollen tube as it grows, the gametes being released near the egg cell when the tip of the tube disintegrates.

What happens next is of critical importance because *both* male gamete nuclei have an important role. One fuses with the egg cell – the moment of fertilization – to form a zygote. This is the first cell of the new sporophyte and it will divide repeatedly to produce the embryo. The second travels to the centre of the female gametophyte and fuses with the polar nucleus to give rise to a unique triploid nucleus containing *three* sets of chromosomes; one from the male gamete and one each from the two female nuclei that fused to form the polar nucleus. From this new triploid nucleus a tissue called the endosperm develops, and it is in the endosperm tissue of maize, wheat, barley, oats, rice and other cereals that the whole of the seed's food reserves are stored. There is not the slightest doubt that endosperm is the single most important source of food for man on this planet: the supermarket bag of flour is, quite literally, ground up endosperm.

Once fertilization has occurred, the ovule technically becomes a seed and the ovary a fruit. Development patterns following fertilization vary greatly from one species to another but essentially three processes take place. First, the zygote or fertilized egg develops into a little plant, an embryo, with recognizable root, shoot and leaves. Second, the cells of the integument develop thick, hard walls in order to form the tough seed coat. Third, a supply of nutrient is transported into the seed by the

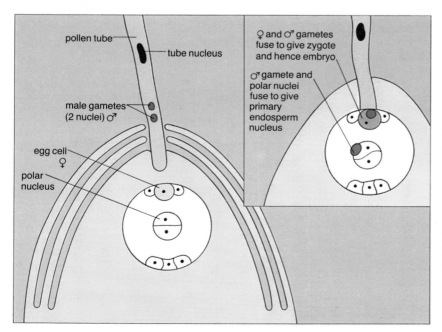

pollen tube
tube nucleus

♀ and ♂ gametes fuse to give zygote and hence embryo

♂ gamete and polar nuclei fuse to give primary endosperm nucleus

male gametes (2 nuclei) ♂

egg cell ♀

polar nucleus

mother plant along the flower stalk and the fruit, so that the embryo has an adequate energy reserve to sustain it during the early stages of germination. This food reserve is located in one of two places, rarely in both. In most dicotyledon seeds it is stored wholly in the first two leaves of the embryo, the cotyledons, which become hugely swollen as in peas and beans. In many, but not all, monocotyledons, but especially in the grasses and cereals, the food reserves are not stored in the embryo at all, but rather in the endosperm – a mass of tissue, packed with storage starch and proteins, in which the embryo becomes embedded.

After fertilization the ovary wall may dry up and disintegrate, releasing the seeds, as it does in peas and beans; or it may develop into a large fleshy structure like a plum or peach. An infinite variety of fruits can arise from the different development of the inner, middle and outer layers of the carpel or ovary wall, while some complex fruits like apples and oranges arise from the fusion of a number of ovaries.

The large, attractive, coloured flowers with which we are all familiar rely, for the most part, on animals to effect their pollination. Insects, birds, bats and other mammals can all be involved in pollination, and some of these relationships are highly specific, neither the plant nor the animal being able to exist without the other.

Other flowering plants such as the grasses and many tree species do not depend on animals, but rely instead on producing large quantities of pollen which is blown by the wind to other plants. In wind-pollinated plants, the flowers are inconspicuous in terms of colour, but they may be large and the stamens may be coloured. What they lack are large, highly coloured, petals and sepals. In trees such as willow and birch, there are separate male and female flowers, in the former on separate trees and in the latter on the same tree. The catkins are the male, pollen-producing flowers and are fairly conspicuous, while the female flowers are rather small. The flowers of grasses have both anthers and ovaries and these are often so small that a lens has to be used to see them. However, they have all the basic structures of a flower.

Both the male and female gametophytes in angiosperms are reduced to a few cells which literally never see the light of day because they are always totally enclosed – by the pollen grain in the case of the male, and by the tissues of the ovule and the ovary wall in the case of the female.

The embryo of the angiosperm seed has to grow and establish itself in a very competitive environment. In its path lie a great many problems, all of which must be solved before it can attain the mature state in which it can reproduce itself. Now that we have examined the plant kingdom and its varied life-styles, we can concentrate on the most sophisticated and important group, the flowering plants, and find out how their growth, development and behaviour are controlled, and how they solve the many problems they encounter.

The male flowers of a wind-pollinated species such as the willow (*Salix*) have no obvious sepals or petals. Their only colour is in their bright yellow anthers and pollen.

The separate male and female flowers of a sweetcorn plant lack colourful sepals and petals and occur on different parts of the plant. The male flower (*far left*) occurs at the apex of the stem and the female (*left*) much lower down. The mass of pink styles can be seen at the end of the female flower. Each is connected to one individual ovule which, when fertilized, will form a grain on the corn 'cob'.

CHAPTER 5

Getting Started

Throughout the warm summer months, fields, hedgerows, meadows and gardens are bright with the colours of flowering plants, and the air is alive with the sounds of insects attracted to them. The flowers then fade, their task completed, to be replaced by slowly maturing seeds and fruits – some bright and conspicuous, others so small and inconspicuous that they are hard to find. The seeds themselves then fall to the ground, or are carried away by the wind, in the guts of birds, or on the coats of passing animals. Finally, as the last leaves fall, winter closes in. For most seeds, this is a time of waiting. But even waiting is an active process. Under the influence of time, temperature, seeping ground-water and its own internal chemistry, the seed is preparing for the next round of activity which, when conditions are just right, will release its stored energy and send its tiny shoot thrusting upwards through the soil and into the sunlight.

The seed's internal structure

Seeds consist of three main parts, two of which are always present while the third may be present or absent according to the species. The protective outer layer, or seed coat, is extremely tough because its cells develop very thick, hard, walls after fertilization has taken place. It is also often very resistant to water and gas penetration, due to impervious waxy substances that are deposited on its outer surface during the final stages of development. The considerable mechanical strength of the seed coat is often one of the factors that prevents germination; in some seeds the growing embryo simply can not at first generate enough internal pressure to burst it open.

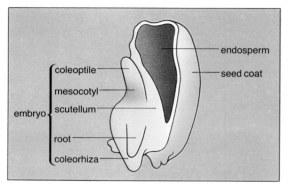

Within the seed coat there is always a young plant – the embryo – although in some species it is not completely formed at the time the seed is released. All embryos consist of the principal structures of the adult plant; a root, a shoot and, joined to this root-shoot axis, one or two embryonic leaves called cotyledons. In some plants these are thin and papery, easily recognizable as leaves because of their shape and arrangement of veins, despite being white through lack of the green pigment chlorophyll. In others they may be very large and swollen, scarcely recognizable as leaves at all. In these cases, they contain the food reserves of the seed.

The optional component of the seed is the endosperm, a tissue which, as we saw in Chapter 4, is very unusual in that it develops from the fusion of three nuclei. Although it is common to classify seeds as endospermic or non-endospermic, there are quite often minute amounts of endosperm in non-endospermic seeds and so the distinction is not quite as absolute as might be thought. The endosperm is, however, an extremely important tissue because in endospermic seeds it contains the

In this striking view of a bisected sweetcorn seed the starch-rich endosperm—the food store of the seed—has been stained blue-black with iodine.

At this stage the pale cream-coloured embryo plant consists of the scutellum, or cotyledon, and the primary shoot and root. The apical bud and tiny young foliage leaves of the shoot are protected by the hollow cone-shaped coleoptile, and the root, too, is protected by a sheath—the coleorhiza.

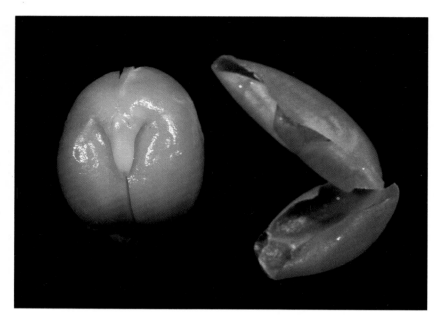

Technically, a seed is said to have germinated when the embryonic root emerges from the seed coat. Most seeds, however, will not germinate immediately they are formed. They usually pass through a period when they are dormant – that is, they will not begin to germinate even in the most favourable conditions. Some months later, however, seeds taken from the same stock will germinate immediately, having lost their dormancy. The ability of seeds to germinate – that is, their viability – usually declines on storage, sometimes very rapidly, but usually rather slowly. Seeds of the rubber tree, the sugar maple and the willow are viable for only a week or so, whereas others appear to retain their viability for several hundred years. Seeds of *Mimosa glomerata*, for example, which had been kept in dry storage in a her-

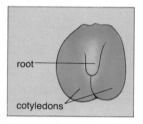

Removing the seed coat from a pea reveals no endosperm, just an embryo with two huge cotyledons.

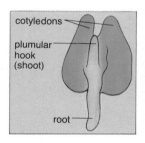

After a few days of germination, the root has lengthened and the shoot, which earlier was tucked between the cotyledons for protection, has also begun to grow and to push the cotyledons apart.

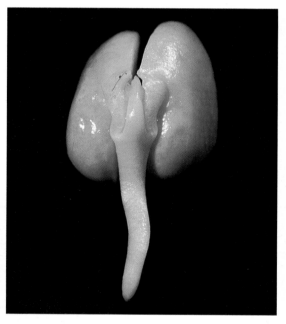

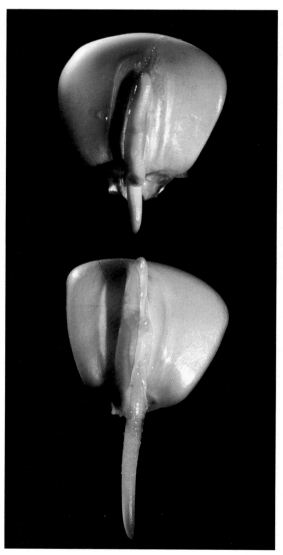

food reserve needed to nourish the embryo in its early stages of growth.

In many plants, particularly the monocotyledons, there are virtually no food reserves in the embryo – they are all stored in the endosperm, which takes up most of the space inside the seed coat and completely surrounds the embryo. In the dicotyledons the endosperm is usually absent as, for example, in pea, bean and sunflower seeds, but there are many dicotyledonous plants in which an endosperm is present and in which the papery-thin cotyledons are embedded, the castor bean and the henbane being examples.

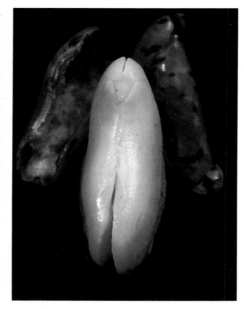

The seed of a runner bean (*Phaseolus coccineus*) seen from the side and end-on (*far left* and *centre left*) gives a clear view of the hilum, the scar where the seed was formerly attached to the fruit, and the micropyle through which water enters the seed to initiate its germination.

Removal of the seed coat (*left*) reveals the radicle, or root, and the two large cotyledons.

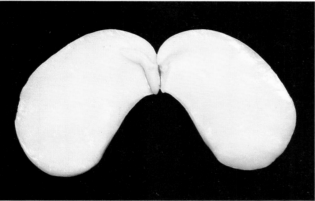

At a much later stage of germination (*below*) the shoot has lengthened to carry the foliage leaves to the surface while the roots, and in this species the cotyledons too, remain beneath the ground.

Separating the cotyledons of a dry seed (above) reveals perfectly formed first foliage leaves and the apical bud of the embryonic shoot. These structures are much more easily observed on seeds that have begun to germinate in darkness (*right*) than on seeds that have germinated in the light.

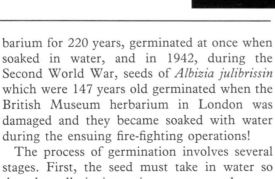

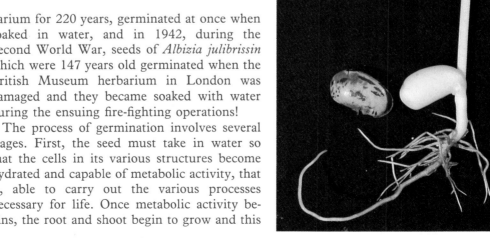

The photographs on the left show two stages in the germination of a maize seed. In the upper picture, 48 hours after the seeds were soaked in water, the primary root has emerged. After a further 24 hours (*below*) the root has lengthened and the shoot (coleoptile) has emerged.

barium for 220 years, germinated at once when soaked in water, and in 1942, during the Second World War, seeds of *Albizia julibrissin* which were 147 years old germinated when the British Museum herbarium in London was damaged and they became soaked with water during the ensuing fire-fighting operations!

The process of germination involves several stages. First, the seed must take in water so that the cells in its various structures become hydrated and capable of metabolic activity, that is, able to carry out the various processes necessary for life. Once metabolic activity begins, the root and shoot begin to grow and this involves the onset of cell elongation and cell division, as well as cell differentiation to provide the specialized tissues needed to carry out certain specific functions. These processes require a great deal of energy, and the embryo can obtain this energy only from its stored food reserves, whether these are held in cotyledons or in endosperm. The second important stage in germination is therefore the mobilization of these nutrients. The third stage is the rapid elongation of the primary root and shoot.

Taking in water
'Dry' seeds usually contain less than 20 per cent water, and most seeds in dry storage have a water content of less than 5 per cent. Some seeds, however, are killed if their water content falls below 30 per cent. These usually fall from the parent plant straight on to moist soil, and are generally characteristic of plants found in the tropics. Mangrove seeds actually germinate

Mangrove seeds germinate while still attached to the 'mother' tree, and develop long green photosynthetically active roots which eventually drop into the mud below to begin an independent existence.

on the mother plant, producing large, downward-pointing, primary roots which grow up to ten centimetres long before being released. These then fall to the ground, penetrating deep into the mud to begin their independent existence.

When a normal dry seed is soaked, it takes up water very rapidly for the first twelve hours or so and then more slowly over a period of several days. The initial rapid uptake of water is due to the hydration of large molecules within the seed so that these molecules and the various structures they make up in the cells of the seed acquire their normal shape and size. This is especially true of the various cell membranes, which once hydrated and reconstituted form the basis for the second, slower phase of water uptake by the process of osmosis, described in Chapter 6. The osmotic uptake of water by the cells of the root and shoot generates the hydraulic pressure required for growth in these organs. Uptake of water by the seed takes place through the minute hole in the seed coat, the micropyle.

Mobilizing the stored food reserves
Regardless of whether they are located in cotyledons or endosperm, the food reserves of the seed consist principally of large molecules such as starch and storage proteins, both of which are largely insoluble in water. Starch usually provides the principal source of energy, while the storage proteins provide amino acids – the essential building bricks which the embryo will use to construct other important proteins. But before the embryo can gain access to this store of energy and building units, the insoluble starch and proteins must be broken down chemically into small units that are soluble in water and which can therefore be absorbed and transported by the embryo.

Just how the nutrient reserves of a seed are mobilized has been worked out most clearly in the endospermic seeds of cereals, particularly in barley, in which the process forms the basis of the brewing, malting and whisky-making industries. The advantage of the endospermic seed in such studies is that the embryo and the endosperm can be physically separated, so that the influence of one on the other can be assessed experimentally.

If a barley seed is cut in half transversely, one half of the seed will contain the embryo and a little endosperm, while the other half contains only endosperm. If each half is soaked separately in water, after a few hours the white starchy endosperm in the embryo half of the seed starts to disappear into a clear solution, while that in the non-embryo half remains unchanged. So, soaking the endosperm in water does not lead to its breakdown; the presence of the embryo appears to be required to initiate the process. If non-embryo half-seeds are placed in a little water, and a number of embryos which have been carefully dissected out of seeds are then added, and the liquid gently stirred, the endosperm will show signs of breakdown within a few hours. Furthermore, breakdown occurs first around the edge of the seed, in the region immediately inside the seed coat.

These experiments indicate that the embryo produces a water-soluble chemical that leads to the endosperm becoming soluble. The simplest explanation for this would be to assume that the embryo produces a chemical substance that directly attacks the endosperm and breaks down the storage starch and protein; but this is not so. If small pieces of endosperm are carefully dissected out of the *centre* of the endosperm tissue and placed in water together with some dissected-out, but quite separate, embryos, no breakdown of endosperm takes place at all. There is, therefore, no direct effect on the endosperm of whatever is being secreted by the embryo. There must be an intermediate process, and the experiments suggest that it occurs in the outermost layers of the endosperm, where breakdown begins, or perhaps in the innermost layers of the seed coat. In fact there is, between the outer layer of the endosperm and the inner layer of the seed coat, another extremely important tissue called the aleurone layer. It is only a few cells thick but it is highly specialized in that its cells alone have the capacity to pick up the chemical signals emitted from the embryo and respond by producing other, quite different, chemicals which cause the breakdown of storage starch and proteins.

The nature of these active substances is now known: they are highly specialized proteins called enzymes, two of the most important being alpha-amylase (α-amylase), which breaks down insoluble starch into soluble glucose, and protease, which breaks down the storage proteins into soluble amino acids. Unravelling the story of how the embryo unlocks the food reserves in the seed has taken a great deal of careful experimental investigation, but the sequence of events is now understood. The

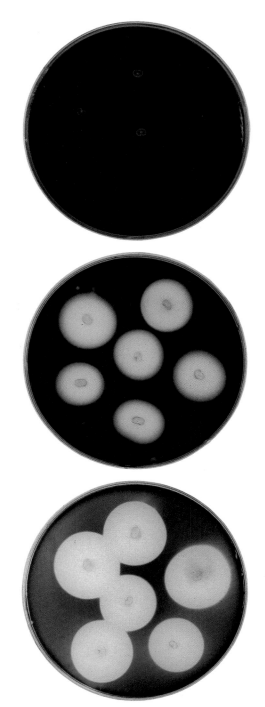

Production of the starch-destroying enzyme alpha-amylase by the aleurone layer lining the seed coat can be demonstrated by treating half-seeds of barley (*Hordeum vulgare*) with gibberellic acid.

The half-seeds used contain no embryo (*see below*). They are placed with the cut surfaces downwards on agar in covered culture dishes. The agar in some dishes contains starch alone, while in others it also contains gibberellin at either low or high concentrations.

The gibberellin first diffuses up into the half-seed where it reaches the aleurone cells and causes them to manufacture and release alpha-amylase. The enzyme then diffuses down into the agar where it breaks down the starch to glucose. The degree of starch breakdown is shown by flooding the dishes with iodine. Remaining starch immediately turns blue-black while clear areas indicate that the starch has been removed.

The dishes shown here contain gibberellin at zero (*top*), low (*centre*) and high (*bottom*) concentrations, and from these results it is clear that the more gibberellin there is in the agar, the more alpha-amylase is produced by the aleurone cells.

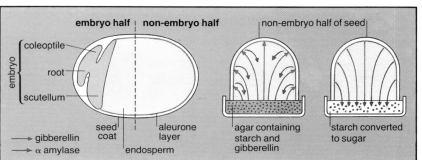

trigger or messenger chemical released by the embryo is a gibberellin – one of a large group of plant hormones (Chapter 16). Gibberellins on their own have no effect on endosperm tissue, starch or protein, but when the specific gibberellin produced by the embryo strikes its target – the cells of the aleurone layer – these cells respond by producing the all-important enzymes, without which the embryo's food reserves would remain beyond reach.

So far we have referred only to the fact that the cells of the aleurone layer begin to *release* the enzymes α-amylase and protease when they are treated with gibberellins. But it might also be useful to know whether or not these enzymes are present in the aleurone layer cells before the gibberellins arrive. The answer to this question will establish how the gibberellins actually operate in an aleurone layer cell. For example, if the α-amylase is already present, then arrival of the gibberellin merely has to bring about its release, perhaps by changing the permeability of the cell membrane. On the other hand, if the enzyme is not present in the aleurone cells before the arrival of the gibberellin, then the gibberellin must initiate the manufacture or synthesis of the enzyme.

A number of experiments have shown that the synthesis hypothesis is correct, and that gibberellin actually initiates the manufacture of enzymes in the aleurone layer cells. The experiments that provided this proof are too complex to describe here, but the way in which plant cells synthesize or manufacture the chemicals they require is a fascinating process and well worth a closer look.

An enzyme is a protein – that is, a long chain of amino acids – that will promote, or catalyse, only one particular chemical reaction. Alpha-amylase will, for example, promote only the breakdown of starch into its component glucose molecules. This very specific nature of an enzyme is determined by the precise sequence in its chain of a number of different amino acids, implying that enzymes have to be made in a very precise way indeed. Information about the required sequence of amino acids is stored on the chromosomes in the nucleus of every cell. A particular part of the chromosome, called a gene, will carry the details of how to make α-amylase in a specific sequence of chemical structures called bases, which are a part of the DNA molecule. In order to make an enzyme, that is, to string together some twenty different amino acids in exactly the

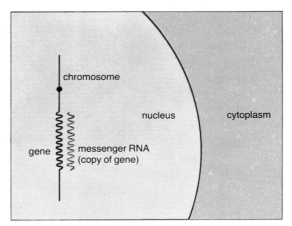

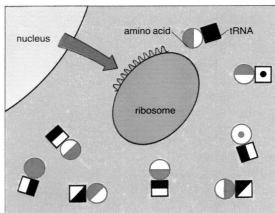

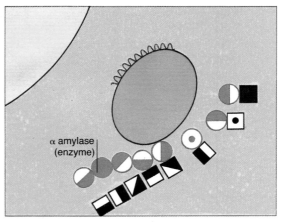

Living cells are complex chemical factories capable of manufacturing a large range of enzymes and other essential compounds.

The genetic code for the sequence of amino acids in any particular enzyme is stored on one short section, or gene, on the chromosomes in the nucleus of the cell.

To construct an enzyme, the cell first makes an exact copy of the required section of code, using a piece of 'messenger RNA'. This messenger carries the coded 'recipe' to one of the ribosomes floating in the cell cytoplasm, and there the required enzyme is built up, step by step, using up to 20 different amino acids collected from the cell cytoplasm by molecular 'shuttles' called transfer-RNA (tRNA).

correct sequence in a chain that may be hundreds of amino acids long, details of the sequence must first be copied on to a molecule that can move from the nucleus, across the nuclear membrane and out into the cytoplasm where the protein synthesis actually takes place. This transfer is achieved by a substance called messenger RNA (mRNA) which is made in the nucleus in such a way that it copies precisely the information in the gene. The messenger RNA then passes out of the nucleus, through the nuclear membrane. On arrival in the cytoplasm, the mRNA operates in close association with the ribosomes, which are small vesicles equipped with the chemical machinery for joining amino acids together. With the arrival of the mRNA blue-print of the required enzyme, the ribosomes are able to string together amino acids in precisely the correct sequence, so that the protein they manufacture has the unique structure that will enable it to catalyse the hydrolysis of starch and so release its component glucose molecules.

The soluble sugars and amino acids released by enzyme action are then absorbed by the embryo and transported to the sites of cell division, growth and differentiation at the tips of its root and stem, and in the leaves. The sugars are broken down by respiration (Chapter 12) to release energy which is utilized to drive other chemical reactions, while the amino acids provide the building bricks for the synthesis of enzymes and other essential components. As the endosperm is broken down during the germination of a cereal seed, the embryo is bathed in a nutrient solution that is more than adequate to supply it with its energy and other needs during the first 7 to 14 days of growth. However, not all enzymes are synthesized in this way in a germinating seed. Some, even in the cereals, may already be present in an inactive form and are activated only in the presence of a protease. Some lipases, enzymes that break down the fats and oils stored in seeds, also occur in an inactive state in ungerminated seeds.

In non-endospermic seeds, where the nutrient reserves are stored in the cotyledons, details of exactly how the reserves are mobilized are less clear. The cells of the cotyledons in the dry seed are filled with starch grains and other small bodies, but these have all disappeared by the time the seeds have been allowed to germinate for 7 to 10 days. There is clearly a marked increase in enzyme activity in the cotyledons, which results in a breakdown of stored nutrients. Enzymes such as α-amylase, protease and lipase are present only at very low levels, if at all, in the dry seeds, but increase markedly once the seeds have been soaked in water. In the case of pea cotyledons, the increase in α-amylase has been shown to be due to enzyme synthesis, a situation similar to that described in cereals.

Exactly how enzyme synthesis is regulated in cotyledons is not fully understood, but it seems that when the seeds are soaked, the uptake of water into the root and shoot of the embryo causes signals to be sent to the cotyledons. This has been deduced from the fact that if the embryonic root and shoot are removed from pea and cucumber embryos, breakdown of the food reserves does not take place and there is no increase in enzyme activity. Also, the fact that an aqueous extract of the embryonic root and shoot will induce enzyme activity in detached pea cotyledons is clear evidence that a water-soluble chemical signal is involved. In peas, however, it seems that this signal is not a gibberellin, because application of synthetic gibberellins does not induce nutrient breakdown. There are, however, many gibberellins (see Chapter 16) and only a few have been tested so far. In the squash, for example, another member of the cucumber family, gibberellin also had no effect, but two members of another group of plant hormones called cytokinins did trigger the breakdown of stored nutrients.

In principle, therefore, the mechanism for mobilizing the stored nutrients in the cotyledons of non-endospermic seeds is similar to that in cereal seeds in that the root-shoot axis releases a chemical signal that induces the production of enzymes in the cotyledon cells. However, the chemical nature of the signals may differ, and whether there are special 'target cells' in the cotyledons, similar to those of the aleurone layer in cereals, has not been established.

Respiration and metabolic activity

Respiration in plants, as in animals, is the process of burning sugar molecules to release their energy so that it can be used to drive other processes. The reaction involves the consumption of oxygen and the production of carbon dioxide, and it can be measured by analysing the rate at which oxygen is taken in by the seed.

In dry seeds, respiration is scarcely detectable. On soaking the seed, however, respiration increases very sharply during the phase of rapid water uptake and then remains constant for some hours, or even days, until the root emerges from the seed. At that point it increases again as the seedling grows and develops. A good supply of glucose from the breakdown of food is necessary to maintain the high rate of respiration required by the seedling in order to build the structural and metabolic systems of its developing organs.

Dormancy

Many seeds kept under dry conditions will germinate as soon as they are exposed to favourable conditions, which usually mean a supply of water, a moderate temperature of 15°C to 25°C and a supply of oxygen. Such seeds are said to be quiescent, rather than dormant.

Dormant seeds are those that even under ideal conditions do not germinate until they have been under those conditions long enough for some changes to have been brought about in their chemical composition, for water to penetrate the seed coat, or for some specific signal to have been received from the environment. In other words, in dormant seeds there is some form of block to the normal process of germination, and that process cannot begin until the blocking mechanism is removed. A number of different blocking mechanisms are known, and their study is of considerable interest to the horticultural and agricultural industries since a prolonged dormant period is highly undesirable in a crop plant. Commercial breeders will, of course, try to breed out prolonged or complicated dormancy mechanisms from crop species. Let us now examine the main types of dormancy and the ways in which they can be broken.

After-ripening

Seeds of many plants, including many crop plants such as corn, peas and beans, show no significant dormancy; they will germinate almost immediately after the seed has matured. Seeds of other plants, however, will not germinate immediately they are mature, but will do so after some weeks or months of dry storage. This natural loss of dormancy is called after-ripening, and is found in many cereals such as wheat and barley. The seed coat seems to be involved in the after-ripening process. In many species it is largely impermeable to gases such as oxygen and carbon dioxide, and the severe limitation of the amount of oxygen available to the tissues inside the seed coat, especially the embryo, may reduce metabolism to a very low level. Removal of the seed coat usually results in immediate growth of the embryo in such dormant seeds, though whether this is due to the enhanced supply of oxygen, the removal of the physical constraint on the developing embryo, or even the removal of inhibitory chemicals present in the seed coat, is a complex question requiring careful examination in each individual plant species.

Immaturity of the embryo

In the seeds of *Anemone*, *Caltha*, and some orchids, the embryo is not fully developed when the seeds are shed; indeed in some orchids the embryo consists of little more than a few cells at this time. No germination is possible in such seeds until the embryo has had time to complete its development.

Impermeability of the seed coat

In a number of plant species such as sweet pea, wood mallow and deadly nightshade, the coat of the freshly-released seed is totally impermeable to water – without which germination is impossible. Such seeds often lie dormant in the soil for years until the seed coat has been rendered permeable by the destructive action of fungi and bacteria. The seed coats can, however, be made permeable to water in commercially important seeds by the process of scarification, that is, by shaking the seeds in a drum with an abrasive medium like sand, or even by removing the waxy layers by soaking the seeds briefly in sulphuric acid.

In some seeds, breaking of dormancy seems to involve the removal, from the seed, of chemicals that inhibit growth and germination. These may be in the seed coat or in the embryo itself. They cannot, however, be removed until the seed coat becomes permeable to water, so that percolating ground-water can wash away the inhibitors. In charlock, which has a potent inhibitor in its seed coat, the process of making the coat permeable, and washing out the inhibitor, can take 10 to 15 years, since the seeds can remain viable in the soil for this length of time. They often germinate after a field has been ploughed when, no doubt, many of the seed coats are damaged and water penetration occurs for the first time.

The need for chilling

It is possible to break dormancy in the seeds of many temperate species by chilling them, that is by exposing them to a temperature of between 0°C and 5°C for several weeks. Common examples of such cold-requiring seeds are apple, peach, rose, pine and birch. In the natural environment this type of dormancy ensures that seeds shed in the autumn do not germinate until the spring. The cold exposure is, however, effective only if the seeds are fully imbibed with water and properly supplied with air, so it seems that some important chemical reactions go on during the chilling period, and that these lead to the release of the seeds from their dormant state. In some species, such as birch and sycamore, removing the seed coat allows the embryos to germinate without any chilling period being necessary, whereas in other species, for example the peach, the embryo will not germinate unless it has been chilled, even when the seed coat has been removed. The imposition of this kind of dormancy is, therefore, a complicated business involving either the seed coat or the embryo, and there are species in which both structures are involved.

What actually happens during the chilling period is not known, but the fact that many such cold-requiring seeds will germinate without being chilled if they are treated with a gibberellin solution, suggests that some change in the level of the seeds' natural gibberellins occurs at temperatures between 0° and 5°C. However, it seems that a variety of chemical substances will substitute for the chilling period – the gas ethylene, substances known as cytokinins (see Chapter 16) and even nitrate salts have all been found to be effective.

Chemical inhibitors of germination can occur in the seed coat, embryo and endosperm of seeds, but their precise role in dormancy has been difficult to establish. The amount of inhibitors in apple seeds appears to decline markedly during the period of chilling, while the amount of growth-promoting substances increases. Dormancy in cold-requiring seeds may therefore be associated in some way with a balance between chemical substances that inhibit and promote germination, and during chilling the balance of these chemicals may change. Even in this one category of dormancy there is such a variation in behaviour from species to species that no simple explanation can be offered.

The need for light

Light can be a potent factor in breaking dormancy and is known to be an important promoter of germination in birch, lettuce, tomato, tobacco, and many other species. Such seeds are said to have a light requirement. Equally, seeds of other species are prevented from germinating by light, examples being the hellebores, deadnettles, campions and catchflies. It is useful to pause here and look more closely at the way in which light can promote germination, since detailed study of this phenomenon in the 'Grand Rapids' variety of lettuce led to the

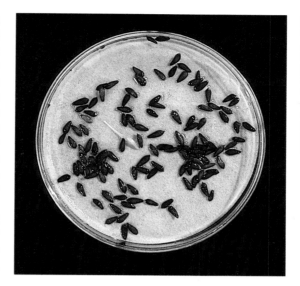

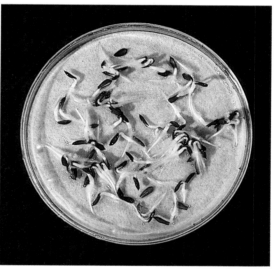

Seeds of the 'Grand Rapids' variety of lettuce (*Lactuca sativa*) do not germinate (*left*) if they are kept damp, at 30°C and in total darkness. However, those same seeds will germinate readily (*below left*) if they are exposed to white light for even just a few minutes. They are said to have a 'light requirement' in order to germinate.

discovery in 1952 of the very important plant pigment phytochrome (Chapter 8), which regulates the growth and development of plants at several points in the course of their lives.

Seeds with a light requirement for germination usually require exposure only to a very weak light for a short time; often a few minutes will suffice. Very little input of radiant energy is therefore required. Lettuce seeds placed on wet blotting paper in a dish will not germinate in darkness, but those exposed to light for less than five minutes show virtually 100 per cent germination. Whenever such a finding is made, the question arises of what pigment molecule is perceiving the light stimulus, because light can have no effect unless it is first absorbed by something. In order to identify the pigment involved, the relative effectiveness of different wavelengths of light must be determined and

When damp lettuce seeds are exposed to light of different colours—that is, different wavelengths, only light in the yellow-red part of the spectrum will trigger germination.

The photographs beneath the action spectrum show the reaction of identical batches of moist seeds. In the dishes exposed to red and yellow light the emergence of the radicles (roots) from the seeds is most clearly visible if the dishes are viewed on a black background.

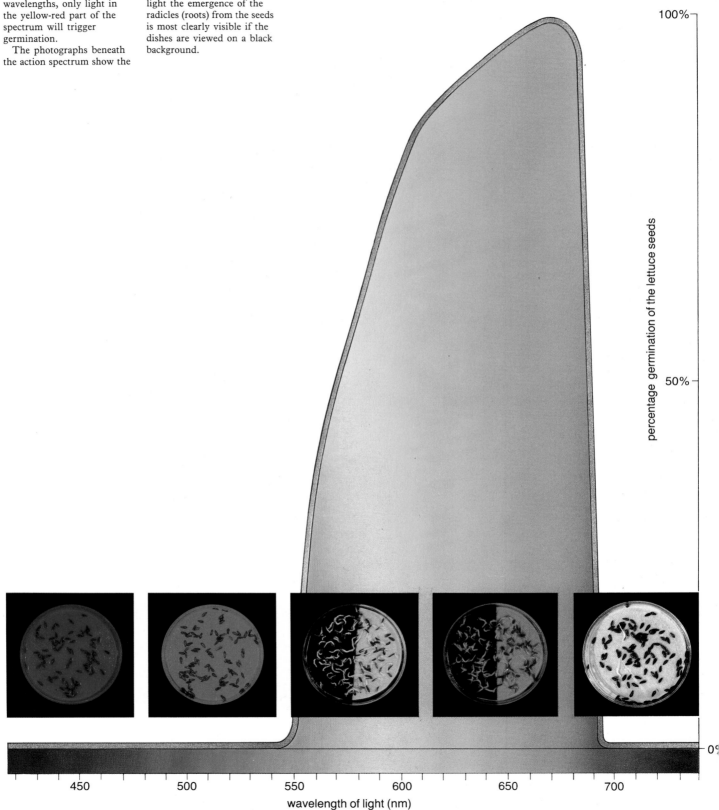

percentage germination of the lettuce seeds

100%

50%

0%

wavelength of light (nm)

450   500   550   600   650   700

compared. In this case, the promotive effect was found to be confined entirely to the yellow-red band of the spectrum, lying between wavelengths of 580nm and 700nm. What was even more surprising was the discovery that if the seeds were given a five-minute exposure to red light and immediately afterwards exposed for five minutes to the far-red spectral band, which lies between 700nm and 800nm, they did not germinate at all! Astonishingly, a further five-minute exposure to red light promoted germination. This phenomenon involves a very precise light-switch mechanism and will be discussed in more detail in Chapter 8. It is the classic example of a plant developmental process controlled by light through the pigment phytochrome.

It is not yet clear how activation of phytochrome breaks dormancy in a light-requiring seed, but there is some evidence that the seed coat has an important role to play since excised embryos from light-requiring birch seeds have no light requirement at all and will germinate quite readily in darkness. Many chemical substances are known to induce germination in light-requiring seeds, and the role of the naturally-occurring growth promoters and inhibitors has been investigated in some detail; but so far no clear picture has emerged. Again, it is likely that some critical balance of promoters and inhibitors may be changed after exposure to red light, but the nature of the change, what chemical substances are involved, and how phytochrome activation achieves such changes are still to be established.

To summarize, then, germination is a complex process in which the principal components are penetration of the seed coat by water, and its uptake by the embryo; the start of metabolic activity in the root and shoot; transmission of chemical signals from the embryo to the aleurone layer or cotyledons to release the seed's food store, and finally the onset of growth.

Dormancy is the suppression of germination by various blocking mechanisms which prevent one or more of these processes from taking place. Removal of the blocks can involve a variety of environmental stimuli, the most common being the effects of water and soil microorganisms on the seed coat, and exposure to low temperatures and to light. However, it must also be remembered that some seeds do not germinate until they have passed through the gut of an animal, usually a bird, or until they have been scorched in a bush fire.

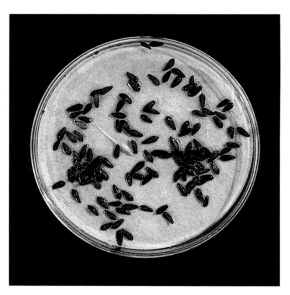

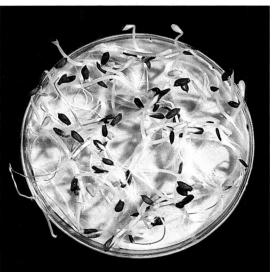

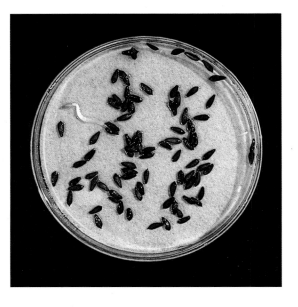

In 'Grand Rapids' lettuce seeds, germination can be 'switched on and off' at will by exposing the seeds to light of two critical wavelengths. Exposure to far-red radiation (700nm to 800nm) completely wipes out the effect of any earlier exposure to red light (600nm to 700nm).

The seeds in these dishes were placed on wet filter paper at 30°C. They were then either kept in continuous darkness (*top*), or exposed to red light for five minutes (*centre*) or exposed to red light for five minutes followed immediately by exposure to far-red radiation for five minutes (*bottom*). The effect of this remarkable internal light switch can be seen in the seeds' response.

CHAPTER 6

Power for Growth

A considerable amount of force is required for the root and shoot of a germinating seed to burst free from the confines of the seed coat and drive their way through the soil. Yet it is only when we see masonry split by a sapling, or concrete paving slabs cracked and lifted by growing tree roots, that we begin to appreciate the remarkable power of plant growth.

These impressive forces are due entirely to the hydraulic pressures that build up inside the individual cells of plant organs, and these pressures are essential for growth, for without their stretching effect on the cell walls, cell enlargement can not take place. This chapter will explain how the internal hydraulic pressure is generated in a cell, how it underlies root and shoot growth, and the vital role it plays in giving strength and rigidity to plants.

Osmosis

To understand how a plant cell generates a large internal hydraulic pressure it is necessary to understand a phenomenon of physical chemistry known as osmosis, for it is upon this process alone that pressure generation depends.

Osmosis arises entirely from what is known as the chemical potential of water. Molecules of water, like those of all other substances, are in constant motion, thrashing about in all directions, and it is the speed of this motion that determines the chemical potential. It is a measure of the free energy of the water molecules – that is, the energy available for doing work – and is, therefore, a measure of their tendency to diffuse, evaporate or be absorbed. Chemical potential is expressed in units of pressure, for example Atmospheres, bars, or in the most recent terminology, Megapascals (1 Mpa = 10 bars = 0.98 Atmospheres).

Pure water can thus be envisaged as a large number of molecules thrashing around in random violent motion, and its chemical potential, or free energy, will have a maximum value at any particular temperature and pressure. The value will increase if the temperature is raised, because the molecular motion will be speeded up, and at boiling point the molecules will escape rapidly from the surface of the liquid in the form of an invisible gas – water vapour. Molecules escape in the same way at a lower rate at room temperature by the process of evaporation. Applying pressure to the water increases its chemical potential because the molecules are squeezed closer together.

The chemical potential of pure water can be reduced in a number of ways – for example by lowering the temperature or pressure, or, most importantly for this discussion, by dissolving other molecules in the water to form a solution. Reducing the temperature slows the motion of the water molecules, reducing the pressure increases the distances between them, and dissolving a solute such as sugar in the water introduces large molecules with which the water molecules collide and to which they may be attracted, thus reducing their activity. So, at a particular temperature and pressure the water molecules in a sugar solution will have a lower chemical potential than those in pure water, and the more concentrated the sugar solution, the lower will be the chemical potential of the water molecules.

If pure water, or a sugar solution, is held in a container, whether it is a laboratory test tube or a plant cell, the container and its contents are called a system, because the various components, including the container, can interact with each other. In any such system we may

Plant growth may be slow but it is enormously powerful. With internal hydraulic pressures of up to 20 Atmospheres in their cells, plant roots can expand with enough force to split concrete, brickwork and masonry.

56

be concerned solely with the chemical potential of the water molecules, or we may wish to consider the chemical potential of the other molecules dissolved in the water. However, in looking at how hydraulic pressure is built up inside plant cells we are concerned only with the chemical potential of the water molecules, and to simplify the terminology, plant physiologists refer to this as the 'water potential' of the system.

We have already seen that the water potential of a sugar solution is less than that of pure water. The speed of motion of its water molecules is therefore less than that of the molecules in pure water. If pure water and a solution of sugar are carefully brought into contact, with no mixing, the water molecules will move from the pure water into the solution more quickly than those in the solution will move into the pure water, so a net movement of water molecules will take place from the pure water into the sugar solution. This process of molecular movement is known as diffusion. The molecules of any substance will always diffuse from a region of higher chemical potential to one of lower chemical potential – that is, from a region of higher concentration to one of lower concentration. In the example above, water molecules diffuse from the pure water, which is 100 per cent water, to the solution, which is less than 100 per cent water. For the same reason sugar molecules diffuse in the opposite direction.

Having established that molecules diffuse along concentration gradients because of differences in their chemical potential, it is easy to understand the process of osmosis because it is really just a special case of diffusion. Osmosis takes place only in the presence of a semipermeable membrane, that is, one that is permeable to water alone. For osmosis to occur, the membrane must separate a solution from its pure solvent, or from a solution of different concentration. The free energy of the water molecules thrashing against one side of the membrane will therefore be higher than that of the water molecules thrashing against the other. This means, in effect, that there is a difference in the pressure being exerted by the water molecules at either end of the microscopic pores in the membrane through which water, and water alone, can pass. The result is that water passes from the side with the higher water potential to the side with the lower water potential. But, most importantly, there is no

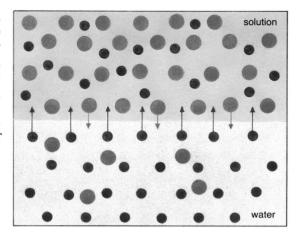

sugar molecule
water molecule

When a solution of sugar is placed very gently in contact with pure water (*left*), sugar molecules start to migrate across the boundary into the pure water while molecules of water migrate in the other direction. Both are moving from regions of high concentration to regions of lower concentration, and this important process is called *diffusion*.

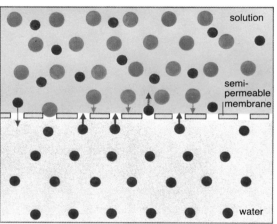

If the same two liquids are separated by a semi-permeable membrane, that is one permeable only to water, we can observe the process by which high hydraulic pressures are generated in plant cells. It is called *osmosis*.

As in diffusion, the molecules try to migrate from high concentration to low – but in this case the large sugar molecules are unable to pass through the pores of the membrane. Water, however, continues to pass through and into the solution beyond.

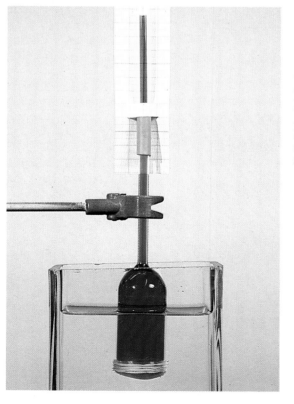

In the photograph below, the upturned funnel has been covered with a semi-permeable membrane, filled with sugar solution (dyed red) and lowered into a beaker of water. At once, the effect of osmosis can be seen as water passes through the membrane and into the solution, pushing the level of the coloured solution up the stem against the pull of gravity.

reverse movement of the larger solute molecules because they can not pass through the pores of the membrane.

Osmosis is easily demonstrated by firmly fixing a semipermeable membrane over the end of a funnel, which is then inverted, filled with sugar solution, and immersed in a beaker of pure water, as illustrated opposite. The water potential of the solution inside the funnel is lower than that of the pure water outside, so water molecules pass into the funnel through the membrane – forcing the liquid to move up the stem of the funnel.

All solutions have what is called an osmotic potential, that is, the capacity to generate an osmotic pressure when separated from their pure solvent, water, by a semipermeable membrane. The osmotic potential of a solution is a measure of the difference between the water potential of the solution and that of pure water, and it can be measured using an instrument called an osmometer. An osmometer is an apparatus that in principle resembles the inverted funnel mentioned above, with the addition of a piston in the stem of the funnel so that a measured pressure can be applied to the sugar solution inside the funnel. To measure the osmotic potential of a solution, a sample is placed inside the funnel. The instrument is then used to measure the pressure that is required on the top of the column to stop osmosis taking place. Increasing the pressure increases the chemical potential of the water molecules, so all we are measuring is the pressure required to increase the chemical potential of the water molecules in the sugar solution to equal that of the pure water outside the funnel. At this pressure, osmosis does not occur because the water potentials of the liquids on either side of the membrane are equal.

Osmosis in plant cells
A plant cell consists of the cytoplasm surrounded first by the plasmalemma and then by the cell wall. The cytoplasm usually contains a large vesicle called the vacuole containing a solution of sugars, mineral salts, amino acids and other chemicals, but for the purpose of this discussion the whole of the cytoplasm, including the vacuole, will be regarded as one rather concentrated internal solution, separated from the water outside the cell by the semipermeable membrane of the plasmalemma.

The cell membrane is extremely thin and elastic, and has the important property of being freely permeable to water molecules and largely impermeable to most other molecules. As we saw in Chapter 2, however, molecules other than water *can* pass through the membrane – but only by invitation, and only at special places associated with the large protein molecules embedded in its structure. They can not simply diffuse through like water. The outer cell wall is rather like a microscopic piece of coconut matting; strong, resistant to stretching, but freely permeable to anything small enough to pass between its strands. It is not, therefore, a barrier to the passage of either water or solute molecules, most of which pass easily through its relatively coarse meshes.

A plant cell is therefore an osmotic system in which the internal solution is separated by a semipermeable membrane from the water outside; and because the water potential of the cell contents is less than that of the water outside, water will enter the cell through the plasmalemma. It is this osmotic entry of water into the cell that generates the cell's internal hydraulic pressure. As water enters, the cell's volume increases and, in the absence of any restraining mechanism, the volume would continue to increase until the solution inside became so dilute that its water potential equalled that of the water outside, which in nature is never pure, but rather a very dilute solution. However, this situation would not be reached until the volume of the cell had increased perhaps to the size of a tennis ball! No cell could stand that degree of expansion because the plasmalemma would rupture and the cell would literally explode.

So why don't plant cells swell up and burst when they are surrounded by pure water? The answer is found in that critically important structure, the cell wall. This strong, flexible, slightly elastic structure surrounds the plasmalemma and its contents, and as water passes into the cell through the plasmalemma, the increase in volume creates an outward pressure on the cell wall. The wall stretches a little, and the volume of the cell increases slightly as the pressure builds up, but soon the limit of wall stretching is reached. Inward movement of water stops, and no further increase in volume takes place. At this point the outward pressure exerted by the cell contents is exactly balanced by the inward pressure being exerted by the cell wall. The increase in pressure inside the cell increases the water potential of the internal solution until it equals that of the water outside the cell. Osmosis is therefore brought to a halt, and a state of equilibrium is reached in which the cell is fully pressurized.

The situation in a fully pressurized plant cell is rather like that in a football. The plasmalemma is equivalent to the rubber bladder – flexible and expandable, but liable to burst if continuously inflated without being confined within the leather outer case. The cell wall is the counterpart of the leather case of the ball – tough and flexible, but resistant to expansion beyond a particular critical size. To carry the analogy one stage further, while the football may represent the main features of a plant cell, the bladder alone is a good model for the typical animal cell, for lacking any strong outer wall, an animal cell is easily ruptured by the osmotic equivalent of over-inflation.

Plasmolysis

Just as water molecules move into a cell that is surrounded by pure water, so water can be made to move out of a cell in the opposite direction. Water moves in when the water potential of the cell contents is less than that of the medium outside the cell. To reverse the water flow, the external medium must have a water potential lower than that inside the cell, and this can easily be arranged by placing the cell in a solution that is more concentrated than the cell contents. Under these conditions water moves out of the cell; the internal pressure is gradually reduced and eventually reaches zero when there is no stretching effect on the cell wall at all. If water continues to be withdrawn, the cytoplasm and the plasmalemma contract still further, and are pulled

If a few drops of human blood are placed in a test tube with Ringer's solution, the red blood cells remain intact since the osmotic potential of this solution is almost the same as that of the natural contents of the blood cells. The intact cells (*right*) disperse light, and so the liquid in the test tube (*far right*) appears cloudy.

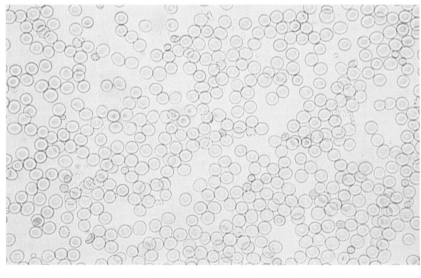

If a few drops of blood are now placed in a test tube with distilled water they quickly swell up and burst, leaving only their 'ghosts' or ruptured cell membranes (*right*) visible under the microscope. The cell membranes are too small to disperse light so the liquid in the tube appears clear (*far right*), albeit slightly pink due to haemaglobin from the burst cells.

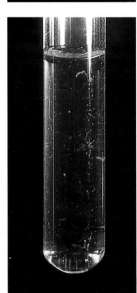

60

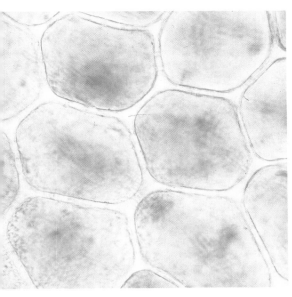

away from the inside of the cell wall. In the epidermal cells of *Rhoeo* leaves, the cytoplasm is brightly coloured and its contraction away from the cell walls is easily observed under the microscope. Cells in which the cytoplasm and plasmalemma have pulled away from the cell wall are said to be plasmolyzed, and the process of water withdrawal by immersion in a concentrated solution is termed plasmolysis.

The contents of plasmolyzed cells exert no pressure on their cell walls, and consequently the cells lose their rigidity. If the cells in a root or shoot are plasmolyzed, the organs become floppy and look as if they have wilted. In fact, wilting is caused by the reduction of the hydraulic pressure in the cells when water is lost by evaporation into a dry atmosphere faster than it can be taken in by the roots. The

In their normal state (*far left*) these cells are surrounded by water. They are fully pressurized and the pink-stained cytoplasm presses against the cell walls But when placed in a strong sugar solution, water is drawn out and the cells become plasmolyzed. First the membrane pulls away from the cell wall (*below left*), then (*below right*) it detaches completely and contracts into a sphere in the middle of the cell.

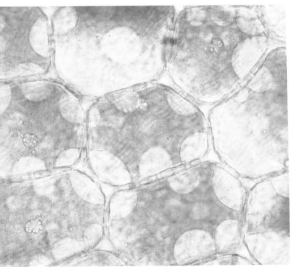

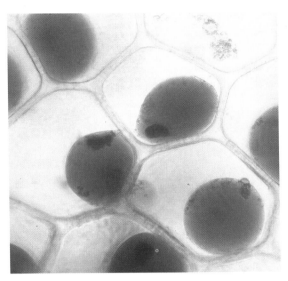

The importance of internal hydraulic pressure. A pot-plant, *Tolmiea*, was allowed to wilt (*bottom left*), its cell pressure reduced to zero, yet 30 minutes after being watered (*below*) it had regained full pressure and could support itself.

mechanism of water withdrawal is different from that of plasmolysis but the end result is exactly the same, and it demonstrates how important the internal hydraulic pressures are, especially in herbaceous plants and young seedlings, in giving the stems and leaves their strength and rigidity. The stem of a young seedling is rather like a hessian sack filled with balloons. If all the balloons are fully inflated they press against one another and against the sack – so stretching it and enabling it to stand up on its own. But if the balloons are only partly inflated they exert little pressure on each other or on the sack, and it collapses. Plants react very quickly to changes in their internal hydraulic pressure, and a wilting plant will often make a complete recovery within half an hour of being watered.

The osmotic potential of plant cells

Quite apart from its scientific interest, the phenomenon of plasmolysis provides a means of actually measuring the pressure that can build up in a plant cell – and hence the pressure that a growing root or shoot can exert on the surrounding soil. The procedure rests on comparing the volumes of the cell cytoplasm in the turgid and plasmolyzed states – and

these volumes can be measured by examining the cells under the microscope. Details of the mathematical calculations that follow need not concern us here, but the results are of considerable interest. In determining the pressure generated in plant cells, values of 10 to 15 Atmospheres are quite common, and these pressures are comparable to those that would be exerted on a body submerged in water to a depth of 90 to 140 metres – the kind of pressure that could crush the hull of a Second World War submarine! In the light of such forces as these, cracked masonry and pavements, and damaged drains, seem a little less surprising, though none the less impressive.

The power for growth

In its normal, fully pressurized state the cell does not grow because the cell wall has been stretched to a point where the inward pressure it exerts on the cell contents exactly balances the outward pressure generated by the osmotic system of the cell. For the cell to enlarge or grow, it has to develop a water potential, that is the capacity to take up more water, and there are two ways in which this could occur. First, the osmotic concentration of the cytoplasm could increase, so raising the internal

These pictures, taken deep underground with the aid of fibre optics, give an idea of the damage and disruption that can be caused by plant roots penetrating underground drains and sewers and then proliferating inside them.

hydraulic pressure in the cell and causing further stretching of the wall. There is, however, no evidence that this happens. Second, there could be a decrease in the tensile strength of the wall, so that it is weakened – with the result that the internal hydraulic pressure of the cell would cause it to stretch still further. This is exactly what happens when a cell grows. The carefully controlled reduction in the tensile strength of cell walls is governed by highly specialized growth hormones which will be discussed in more detail in Chapter 16. The points being emphasized in this chapter are that the driving force for growth in a plant cell arises entirely from osmosis, and that the actual expansion of a cell results from a disturbance of its pressure equilibrium by a carefully controlled weakening of its wall.

The roots and shoots of young seedlings can grow at rates of 0.2mm to 1.0mm per hour. Forward movement of their tips is therefore relatively slow, and the growth of the organs resembles a hydraulic ram in that the apical part, the 'nose-cone', is forced forward by cell expansion in the sub-apical zone pushing against a firmly anchored base – the root mass. Lateral growth, which occurs much later and results in the enormously thick roots and shoots

of trees and shrubs, also depends upon osmosis for its driving force.

The force that a growing root or shoot exerts on fairly loose soil is approximately equivalent to the water potential of the cells in its growing zone. In a medium such as compacted soil, or in crevices between rocks and stones, enlargement of the growing cells may be resisted both by the cell walls and by the medium surrounding the organ. If the medium is extremely resistant, it is possible that most, if not all, of the hydraulic pressure generated in the cells is being held by the surrounding medium rather than by the cell walls. In this case the organ is exerting the maximum possible pressure on its surroundings, a pressure equal to the osmotic potential of the cell contents. So, the structural damage that plants can inflict on roadways and buildings is due to their ability to generate hydraulic pressures of 15 or even 20 Atmospheres, and to maintain those pressures over very long periods of time. Once a root penetrates a crack or crevice in masonry or in drainage pipework its continued growth forces apart the structure causing serious damage. In pipework especially, once penetration has been achieved, the roots can soon cause total blockage by their prolific growth.

In order to study the growth pattern of its roots, this bean seed was germinated against a sheet of glass. The first few days were characterized by rapid and deep penetration of the primary root (*far left*). This was followed two days later by the development of lateral roots (*centre*) and during the following two weeks by extensive penetration of the soil by a mass of root growth (*left*).

CHAPTER 7

Guidance Systems

When a seed begins to germinate, its first and most critical task is to ensure that its rapidly growing root and shoot are heading in the right directions. If the seedling is to have any chance of survival its primary root must reach a reliable supply of water and minerals as quickly as possible. It must also establish a firm anchorage to support the shoot. These objectives are best achieved if the direction of growth is vertically downwards. On the other hand, the tip of the shoot must reach the soil surface as quickly as possible so that exposure to the sun's rays will make the leaves expand and begin to manufacture their own food before the slender reserves of the seed are exhausted. The shortest path to the surface is usually vertically upwards.

The problem of root and shoot orientation is the same whether the seed germinates on the soil surface, where it may have fallen naturally, or buried beneath the soil surface, where it may have been placed by a farmer or gardener, or by the trampling feet of passing animals. Whatever the initial orientation of a seed sown a few centimetres below the soil surface (or indeed of one suspended in humid air in darkness in a laboratory) the primary root and shoot will always direct their growth vertically. This raises the question of what environmental signal the organs use to achieve such a precise alignment of their growth. The signal used is gravity, and we now know that both the primary root and primary shoot of a seedling are equipped with a highly sophisticated gravity-sensing guidance system. This remarkable system is responsible for the phenomenon known as geotropism, which enables roots and shoots to correct their direc-

tion of growth if they are displaced from the vertical.

On emerging from the soil, a shoot may find itself beneath an overhanging rock, or heavily shaded by an already established plant. In the first case, continued vertical growth would lead to a collision, causing physical damage to the tip of the shoot. In the second, it would lead the apex into even heavier shade, with the result that effective photosynthesis would be impossible. When a shoot finds itself unevenly

The bizarre shapes of these coastal mangroves hold a clue to one of the marvels of plant biology. The stilt-like roots do not simply droop towards the water and food-rich mud below: each one behaves like a targeted missile, driven downwards under the control of a highly sophisticated gravity-sensing guidance system.

Despite the steepness of the hillside on which they were planted, these conifers have grown tall and straight and perfectly upright. Their uniform vertical trunks show the precision of the mechanisms that controlled their growth.

illuminated, as it would in such situations, it bends towards the source of brightest light. This phenomenon, known as phototropism, will be familar to readers who grow pot plants on their window sills, and it demonstrates that shoots also possess a light-sensing guidance system.

It is clear then that plants possess two guidance systems, one sensing gravity and the other light. The two operate simultaneously, and interact to control the plant's final angle of growth. Surprising as it may seem, no one – so far – has ever observed or studied a pure phototropic response, because as soon as a shoot deviates from the vertical due to phototropism, its gravity-sensing system is activated and tries to restore the vertical orientation. The final orientation, usually called a 'phototropic response', is therefore a compromise between the two guidance systems acting against each other! Only recently has the exciting possibility of studying a pure phototropic response been presented by the Space-lab programme initiated by NASA and the European Space Agency. In Space-lab there is no effective gravity, and only in such conditions will the true nature of phototropism finally be revealed. Roots, for the most part, appear to lack a light-sensing guidance system. After all, they usually grow in soil in total darkness, and therefore have no need for a phototropic response system. However, we should wait until they have been tested for a phototropic response in the absence of gravity in case their geotropic response totally masks their response to light!

What is a guidance system?

In simple terms a guidance system is a mechanism designed to ensure that a body moving through space keeps to a pre-arranged course. In this respect there is little difference between a root or shoot growing on a predetermined course with respect to gravity, and that of a space rocket proceeding towards some distant galactic objective.

To be effective, a guidance system must have three principal components; a sensing mechanism to detect deviation from the predetermined course, a response mechanism to restore the projectile to the correct course if deviation has been detected, and a mechanism to connect the sensing and response mechanisms, since in both rockets and plant organs they are located some distance apart. Shoots and roots can therefore be regarded, quite accurately, as projectiles proceeding on predetermined courses that have been selected during the process of evolution as those offering the seedling the greatest competitive advantage in the fight for survival.

In order to appreciate the operation of the guidance systems in plant organs it is necessary to understand that the part of the organ in which the component cells are actually growing is not located at the very tip of the root or shoot, but rather some two to five millimetres behind the tip. The very apex of the organ – the 'nose-cone' of the projectile – is where cell division is taking place, so providing a continuous supply of new cells which then undergo elongation (growth) in the sub-apical zone. Here they can triple their length. This zone of cell growth has the effect of thrusting the tip of the organ through the soil, and it is the continuous production and extension of cells in the tip and sub-apical parts of the organ respectively that cause the growth of the organ itself. If the cells in the growing zone extend uniformly then the organ will grow straight, but if they extend more on one side than the other then the organ will develop a curvature. The precise control of the growth rate of cells on opposite sides of an organ is therefore the essential element in the plant's guidance system. Just as the course of a rocket is adjusted by the firing of small thrusters on the sides of the main rocket, so the sensing system in a plant organ can initiate a little more or a little less growth of the cells on one or other side of the organ to alter its angle of growth and restore it to its proper path.

Photographs taken during the first German Spacelab mission in 1985 reveal the behaviour of lentil roots growing in a weightless environment. Photo A shows the roots growing in all directions in the absence of gravity. In photo B the roots are growing uniformly 'downwards' after being placed in a centrifuge which subjected them to an artificial gravity equivalent to that of earth. Photos C and D compare roots kept in weightless conditions with others subjected to three hours of artificial gravity. Even in that short time, the latter have started to change their direction of growth.

SPACELAB D1 MISSION
30.10.85 – 6.11.85
EXPERIMENT 39F : ROOTS

A μg

B 1g centrifuge

C μg

D 1g for 3h

The gravity-sensing guidance system

The curvature developed by a plant organ when it is displaced from the vertical is an active process not a passive one. Roots *grow* downwards after being placed horizontally; they do not simply bend down under their own weight. If a seed is held firmly with its emerging root projecting horizontally just above the surface of a dense liquid such as mercury, the root will grow down into the mercury, despite considerable resistance. By contrast, a root held horizontally in humid air with its root cap removed will continue to grow horizontally for many hours. The root cap contains the gravity-sensing mechanism, and without it the root can not detect that it is growing in the wrong direction.

Only the first, or primary, root and shoot grow vertically; other organs which develop later adopt other predetermined angles to the vertical. The lateral roots, which grow out of the primary root, adopt an angle of approximately 45° from the vertical, even before they emerge from the mother root. The secondary lateral roots which develop from these first laterals grow in all directions. The same is true for shoots; the first laterals on the main stem grow at 45° to the vertical while the secondary laterals are indifferent to gravity. Other plant organs, such as the stems of couch grass and ground elder, grow horizontally beneath the soil surface. These rhizomes have a very precise direction control mechanism which keeps their apices travelling horizontally. Should the tip break the surface, for example by growing out

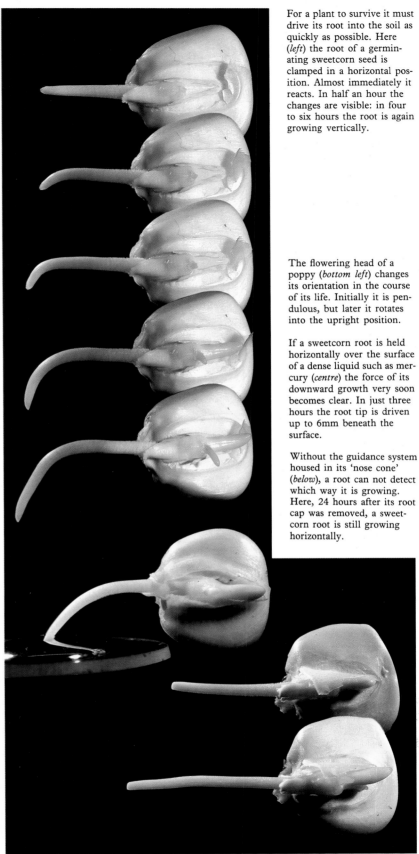

For a plant to survive it must drive its root into the soil as quickly as possible. Here (*left*) the root of a germinating sweetcorn seed is clamped in a horizontal position. Almost immediately it reacts. In half an hour the changes are visible: in four to six hours the root is again growing vertically.

The flowering head of a poppy (*bottom left*) changes its orientation in the course of its life. Initially it is pendulous, but later it rotates into the upright position.

If a sweetcorn root is held horizontally over the surface of a dense liquid such as mercury (*centre*) the force of its downward growth very soon becomes clear. In just three hours the root tip is driven up to 6mm beneath the surface.

Without the guidance system housed in its 'nose cone' (*below*), a root can not detect which way it is growing. Here, 24 hours after its root cap was removed, a sweetcorn root is still growing horizontally.

through the side of a slope, a light-activated guidance system sends it diving below ground again.

We know little about the mechanisms by which plant organs orientate themselves at angles other than the vertical. Virtually all our knowledge about gravity-sensing guidance systems comes from study of the vertically-orientated primary roots and shoots. We will therefore look first at just how plant organs detect their orientation with respect to gravity and then at how they modify the rate of growth of their cells in order to make adjustments to their direction of growth.

The sensing mechanism

Before tackling the gravity-sensing mechanism in plants it is helpful to consider the way in which animals have solved this problem. Most have evolved a relatively large multicellular organ within which there is a cavity containing fluid, and through this fluid a number of heavy particles sink to the lowermost side. There they make contact with sensory hairs on the surface of the cells lining the cavity. Electrical impulses are transmitted along nerves to the brain, whence appropriate instructions are sent to the limbs to make whatever positional corrections are necessary.

In man this organ forms part of the inner ear, and the sedimentary particles are tiny grains containing calcium carbonate. Some crustaceans, such as the crayfish, have a sensory chamber that is open to the exterior. Early in the animal's life, sand is accumulated in this chamber where it combines with a jelly-like substance to provide a heavy mass which normally rests on the lowermost side of the chamber. Any change in the pressure there, caused by tilting the animal, produces the leg action required to correct the tilt. The critical role played by the sand grains can be illustrated by allowing young crayfish access only to iron filings rather than sand. These become incorporated into the gelatinous mass and press downward quite normally when the animal is upright. But if a magnet is placed above the animal it will attract the mass, raising it and making it press on the uppermost surface of the cavity, as it would if the animal were upside down. Despite being the right way up already, the animal will try to turn over.

Animals, then, appear to perceive gravity by means of an organ containing one or more heavy bodies, called statoliths, which sink to the bottom of a fluid-filled cavity. The question is, do plants use a similar mechanism or have they developed something quite different?

Large multicellular chambers like the human inner ear or the sensory organ of the crayfish are unknown in the plant kingdom. But in 1900 it was discovered that all gravity-sensitive plant organs possessed individual *cells* in which minute particles appeared always to lie against the lowermost wall. In the root these cells, the statocytes, are very localized. They occur only in the root cap. In stems they are distributed all along the organ, and usually occur in rings around the vascular bundles. Because they are so localized, the root statocytes are much easier to study than those in stems, and the electron microscope has revealed that the sedimentary bodies in the statocytes are, in fact, clusters of starch grains surrounded by membranes, rather like small plastic bags full of marbles. These structures are called amyloplasts. In principle, therefore, plants appear to have adopted a similar solution to animals for the problem of gravity detection, but the scale is very different as the plant system operates within individual cells. But can we be sure that the sedimentable amyloplasts really do function as a gravity detector in plants?

Several observations support this view. First, all organs that respond to gravity have sedimentable amyloplasts. Second, the amyloplasts sink to the lowermost side of the statocyte cells in less time than the organ takes to detect its displacement from the vertical. If this were not so they could not possibly be part of the gravity detection mechanism. It can also be shown that the quicker the amyloplasts sink, the quicker the plant organs respond to gravity. Third, amyloplasts can be removed from a plant root or shoot and when this is done the organ loses its sensitivity to gravity. We have already seen that removing the cap, to which the statocytes are confined, from a root causes it to lose its ability to recognize its orientation with respect to gravity. It is also possible to leave the cap in place and to remove the amyloplasts by chemical means. When this is done the roots again loose their ability to detect gravity, and will grow in any direction. When the chemical treatment is stopped, the amyloplasts reappear, and the roots simultaneously regain their capacity to detect gravity.

Having established that the sedimenting amyloplasts form the first stage in the plant's gravity-sensing mechanism, we need to con-

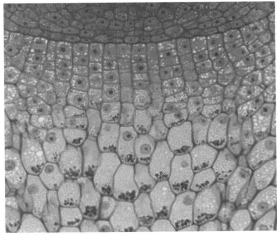

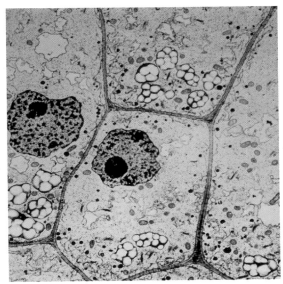

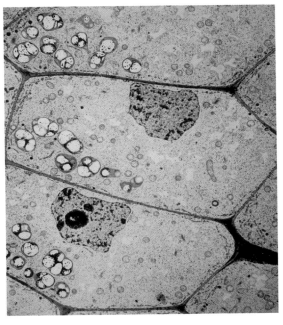

Modern microscopes can take us right inside the plant's guidance system. The photographs above show the external and internal features of a sweetcorn root, with the gravity-sensing cells right at the very tip, inside the translucent root cap.

The close-up (*top right*) of the root cap (light blue) shows the gravity-sensing cells and their clusters of starch grains.

Electron microscope views of a vertical root (*centre right*) and a horizontal root (*bottom right*) clearly reveal the dense starch grains lying on the lowermost cell walls of the gravity-sensing cells.

sider what is actually achieved by the sedimentation of these bodies at the cellular level. The sedimentation in fact sets up an asymmetry in the organ, and as the plane of symmetry is vertical the organ will at once be provided with information indicating in which direction it must curve in order to regain the vertical. Electron microscope studies of the lowermost sides of these cells have revealed no specialized sensory structures capable of detecting the sedimented amyloplasts. However, the amyloplasts do carry an electric charge, and this may affect the characteristics of the outer membrane of the cell on its lowermost side, making this part of the membrane somewhat different from that of the uppermost side of the cell immediately beneath it. This difference between adjacent cells may lead to the one-way, or polar, movement of special growth-regulating chemical substances from the upper to the lower cell.

Distribution of the growth promoter auxin in vertical and horizontal coleoptile tips can be assessed by collecting the chemical in small agar blocks (*top and bottom left*). The amount of auxin in each block is gauged from the degree of curvature it induces when placed asymmetrically on a detipped coleoptile.

The results show that auxin is evenly distributed in a vertical tip, while the lower side of a horizontal tip has a higher concentration than the upper side.

The response mechanism in shoots

As soon as the root or shoot detects that it has been displaced from the vertical, growth-regulating processes are initiated which will restore it to its correct course. The most detailed studies of this response mechanism have been made with the apical part of the shoot of grass and cereal seedlings – an organ known as a coleoptile. This is a hollow sheath, like the finger of a glove, which protects the delicate stem apex and the young foliage leaves, and it is exceedingly sensitive to both gravity and light. Careful observation of the changes in the rate of growth of the upper and lower surfaces of a horizontally-placed oat coleoptile revealed that the growth rate of the upper side was reduced while that of the lower side was increased by approximately the same amount. This observation suggested that a growth-regulating hormone was being redistributed from the upper side of the organ to the lower side, or vice versa.

In the late 1920s it was found that the extreme tip of a coleoptile produced a growth-promoting chemical. It was called auxin. In a vertical organ this substance is transported back from the tip into the sub-apical zone and is evenly distributed among the cells of this zone. The result is that the cells elongate at an equal rate and the organ grows straight. In 1926 it was proposed that adjustment of the growth rates of the upper and lower sides of a horizontally-placed coleoptile was achieved by auxin becoming asymmetrically distributed in favour of the lower half of the organ, and that this asymmetry was due to the gravity-sensing mechanism initiating a lateral transport of auxin from the upper to the lower half of the organ. Since auxin promotes growth, the lower side would grow faster and the upper side slower, thus giving rise to an upward curvature. When the apex regained the vertical position, the gravity-sensing mechanism would 'switch off' the lateral transportation of the growth hormone and the organ would once again grow straight upwards.

Proving this hypothesis correct has taken fifty years and a great deal of effort, but even now it is still not clear whether it is the whole story, or even if the theory is applicable to all types of shoot. First it was shown that an asymmetry of growth-promoting activity did indeed diffuse from the basal end of a detached horizontal

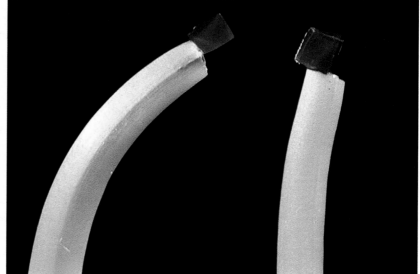

coleoptile tip. Establishing that this asymmetry occurred despite the identity of the chemical or chemicals involved being unknown was a remarkable achievement in itself, but it posed two further questions; first, what was the chemical responsible for growth promotion, and second, did it – or did it not – undergo lateral transport from the upper half to the lower half of a horizontal coleoptile? It will be appreciated that the early experiment shown here revealed only that an asymmetry in growth-promoting activity was set up in horizontal coleoptiles: it gave no indication of *how* the asymmetry was established. That the asymmetry did involve the lateral movement of a growth-regulating compound seemed clear from a simple experiment in which thin slivers of mica were inserted into the tip of a coleoptile either vertically or horizontally. Only the latter would prevent the lateral movement of compounds across the organ, and if such movement was involved in the development of geotropic curvature then only in this case should curvature be abolished. This is what was found. But although this experiment shows that lateral transportation of growth-regulating chemicals is involved in geotropism, it cannot establish whether this involves the upward movement of a growth inhibitor, the downward movement of a growth promoter, or both! The *net* growth-promoting activity measured by the coleoptile curvature test is the overall effect of all the promoting and inhibiting compounds that diffuse out of the apex of the coleoptile, and not a measure of the amount of auxin!

The chemical identity of auxin, the growth-promoting compound present in the coleoptile tip, was only established unequivocally in 1972. Using the analytical technique of mass-spectrometry it was shown that auxin was, in fact, indole-3-acetic acid (IAA). This substance was known to enhance the growth of cells in coleoptiles and other shoots at concentrations as low as just a few parts per million, and its redistribution within the coleoptile thus clearly offered the possibility of adjusting the growth rate on the two sides of the horizontal organ to elicit curvature. About a decade earlier it had been shown beyond doubt that indole-3-acetic acid, which was radioactive and could therefore be tracked within the plant, did indeed move laterally from the upper to the lower half of a coleoptile placed in a horizontal position. It is important, however, to realize that IAA does not simply move downward across a horizontal

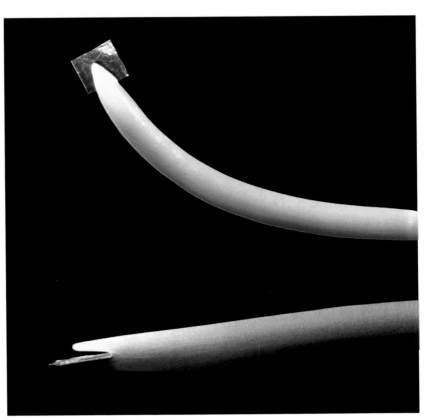

coleoptile under the influence of gravity: the IAA molecule itself is much too small to be directly affected by this force. What happens is that gravity is detected by the gravity-sensing mechanism and then a one-way transport system, specifically 'tuned' to IAA, is switched on. That this is an active transport system requiring an input of energy is shown by the fact that it does not occur if the coleoptile is deprived of oxygen.

Whether or not this sequence of events takes place in the young growing shoots of broad-leafed seedlings is still uncertain. In such shoots the changes in growth rate of the upper and lower sides are not really consistent with the simple redistribution of one growth-regulating chemical. In some species, for example, the upper side stops growing altogether and there is little change, or just a slight acceleration, in the growth of the lower side.

The response mechanism in roots
In roots, of course, displacement from the vertical leads to a *downward* curvature. The cells on the lower side must therefore grow less rapidly than those of the upper side. There are many ways in which this differential growth can be achieved; the growth of the upper and

Inserting thin slivers of metal foil vertically and horizontally into the tips of horizontal coleoptiles proves conclusively that the upward curvature of such organs is caused by the movement of growth-regulating chemicals either upwards or downwards across the organ. The vertical barrier has no effect, but the horizontal one blocks the movement of the chemicals and so prevents the shoot from developing a curvature.

lower sides may simultaneously increase and decrease respectively; that of the upper side alone may increase without a change in the lower side; or the lower side might stop growing altogether, without any change in the growth of the upper side. What actually happens is something of a mystery because different groups of scientists have reported that even in different varieties of the same species, *all* the changes mentioned above can be observed! Roots are rather difficult organs to study experimentally so for the present we must be content with the rather primitive state of our knowledge.

There is no doubt that at least one chemical that inhibits cell growth is produced in the root cap, and that this travels through the root apex and into the growing zone where rapid cell extension is taking place. Its presence has been demonstrated by simple experiments in which either one half of the root cap is removed or in which a small glass sliver is inserted into one side of an intact root. These procedures quickly lead to the root developing a very strong curvature, in the first case towards the side on which the remaining half-cap is located, and in the second, away from the side into which the barrier is inserted. Both experiments demonstrate clearly that at least one chemical substance which inhibits cell growth comes from the cap, travels back along the root and depresses its growth. If this inhibitory chemical is involved in the development of geotropic curvature then it must become asymmetrically distributed when the gravity-sensing mechanism is activated. This asymmetry could be achieved in a number of ways, for example by a lateral transport mechanism moving the substance from the upper half to the lower half of the cap, from which point the asymmetrically distributed substance will then be transported back into the growing zone to give rise to differential growth, and hence curvature. That such a downward-lateral movement of an inhibitor does take place can be shown by the simple experiment of inserting a vertical or horizontal barrier into the root cap as shown in the accompanying illustrations. The mechanism underlying the development of geotropic curvature in roots thus appears to be similar to that in shoots, the major difference being that in roots the chemical involved is an inhibitor of growth rather than a promoter. The chemical identity of this inhibitor is, however, quite unknown.

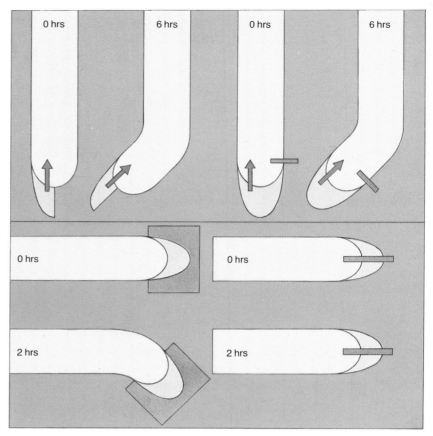

The 'knock-down' response in plant stems

By the time a plant stem reaches maturity, all its constituent cells will have reached their predetermined size and will have stopped growing. There should thus be no potential for further growth, and therefore no possibility of differential growth producing curvature in stems knocked out of the vertical. It follows, then, that without some additional mechanism, mature shoots knocked down by wind or rain, or trampled under-foot by animals, would have no way of righting themselves. Flower and seed formation might never take place, or at best might do so in extremely disadvantageous circumstances and with little likelihood of success.

Many plants do have such recovery mechanisms, one of the best-studied being that found in the mature stems of grasses, and cereals such as wheat, oats and sweetcorn (maize) – some of the world's most important crop plants. Such stems have small swellings at intervals along their length. They occur at what are known as nodes, each node marking the point at which a leaf is attached to the stem. In cereals this structure is a little complicated since the leaf forms a tight, strong, cylindrical sheath around the stem for a considerable dis-

If one half of a root cap is removed (*top left*), the root curves towards the side with the remaining half-cap. If a piece of metal foil is inserted into the side of an intact root (*top right*), the root will bend away from the side containing the barrier. Both experiments show that the bending response is controlled by an inhibitor travelling from the root cap back into the growing zone.

In the lower diagrams, the insertion of a vertical barrier (*left*) has no effect, while the insertion of a horizontal one (*right*) prevents curvature – clear evidence that the growth regulator must be moving either upwards or downwards across the horizontally-held root.

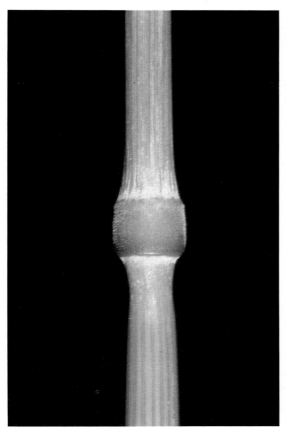

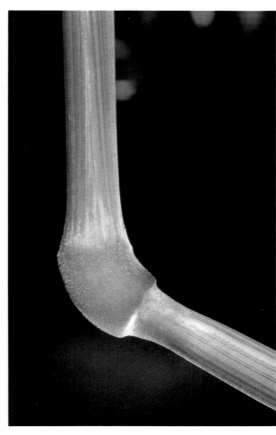

As long as a grass or cereal stem remains upright, there will be no growth at all at the swollen leaf sheath bases or nodes (*far left*). However, if the stem is knocked over the cells on the lower side of the node will start to grow, producing the familiar 'elbow' bend (*left*) that will return the upper stem and ear to the normal upright position and allow the ear to ripen.

tance before it separates to form the familiar long thin grass or cereal leaf. The leaf sheath is very important as it provides much of the shoot's structural strength. The real stem inside the sheath is in fact quite weak and flexible. Right at the bottom of each leaf sheath is the swollen base, or node, and it is these structures that have the potential for further growth – when, and only when, the stem is knocked over.

If a stem remains upright the growth potential of its nodes will never be called upon, but if it is knocked down the recovery mechanism is activated and growth begins immediately on the lowermost side of each node in the horizontal section of the stem. Within a few days the stem regains the upright position by developing a series of sharp bends at the nodes. Although deformed, the stem is once again able to perform its essential function of holding up its leaves and ear to the rays of the sun so that the seeds may ripen.

So how does the node work? First, it contains a gravity-sensing mechanism involving amyloplast sedimentation just as we saw in the root cap. In this case the amyloplasts come to lie against the outermost walls of the statocytes –

that is, the walls next to the outside of the shoot – and they effectively switch on the growth of the cells adjacent to them; cells which otherwise do not grow at all. The statocytes in stem nodes must be rather more sophisticated than those in roots and young shoots because the statoliths can lie against virtually any part of the cell *except* that adjacent to the outermost wall without initiating growth. What is special about the outermost walls of these cells is not yet known, but they are certainly the active or sensitive parts of the cells. It is easy to cut the node out of a stem and then divide it lengthways into two pieces. If both halves are placed horizontally, one with the outer surface upwards and the other with this surface downwards, only the latter grows. The observation also reveals another important characteristic of the response of a grass or cereal node. The growth of the lower half does not depend on the movement of anything from the upper half of the node. The induced growth is highly localized and does not depend on chemical substances coming from anywhere else. Unfortunately we still do not know exactly how the localized growth response is switched on and off.

Light-sensing guidance systems

The light-sensing, or phototropic, guidance system in plant shoots is probably as sensitive to visible radiation as the human eye. As with all sensory systems, the stimulus – light – has to be perceived before any reaction is possible, and the only way light can be perceived is for it to be absorbed by a chemical substance. The energy gained during the absorption process is converted into chemical energy which is then used to activate other processes. You, the reader, can see these pages only because light is being reflected by the non-black parts of the page into your eyes where it is absorbed by rhodopsin, the black pigment of the retina. The excited rhodopsin then releases its newly absorbed energy, and in doing so starts the processes that eventually send electrical signals along the optic nerves to the brain where the information is interpreted as a picture.

So, the light-sensing guidance system in a plant shoot has to consist of at least two components, a light-sensing mechanism that actually receives the light stimulus and transforms it into an electrical or chemical signal, and a response mechanism which regulates the growth of the stem so that curvature can be produced.

The light-sensing mechanism

Animals and plants differ quite dramatically in their light-sensing mechanisms. Man, for example, has just one mechanism – based on the eye pigment system. Plants, by contrast, have at least three quite distinct photo-receptor mechanisms involving three separate groups of

Like a miniature radar station the lesser celandine flower (*Ranunculus ficaria*) tracks the sun across the sky in a classic phototropic response.

pigments each of which has a different role. These are the phototropic pigments, about to be discussed in detail; phytochrome, which controls the plant's developmental sequence (Chapter 8); and the photosynthetic pigment complex (Chapter 9) which includes chlorophyll and other yellow substances known as carotenoids. Other pigments have no critical energy-capturing or regulating roles; they are primarily of cosmetic importance and include the pigments that provide the red, pink, blue and purple colours in flowers and other plant structures. Their main function is to make the flowers attractive to insects or birds to ensure that the pollination stage of sexual reproduction takes place.

The first task in exploring the light-sensing guidance system in plant shoots is to identify

After more than a century, Charles Darwin's experiments of 1880 remain one of the most effective ways of demonstrating the phototropic response.

Here, the experiments have been photographed using red light as this does not cause a phototropic reaction.

In the upper photograph a sweetcorn coleoptile curves towards the white light of a candle. In the middle photograph, the top 2mm of the coleoptile have been covered with a lightproof foil cap. The shoot does not bend, even though its growing zone is brightly illuminated. In the lower photograph, a screen ensures that the tip alone is illuminated while the growing zone remains in deep shadow. The shoot bends towards the candle – proving that a message must have been passed from the light-sensitive tip to the growing zone farther down.

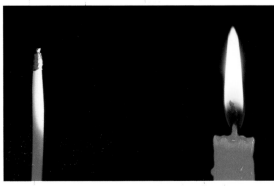

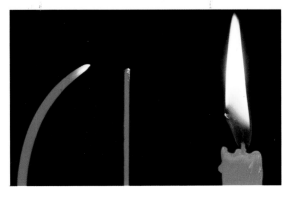

the pigment that is used to detect the light stimulus. The procedure is theoretically quite simple; first, determine the wavelengths of light that elicit the phototropic response, and then grind up the plant and identify the chemical substance that absorbs exactly the same wavelengths. In other words, look for a molecule with an absorption spectrum that exactly matches the action spectrum. In practice, however, this is a time-consuming exercise and one fraught with technical difficulties.

Just to emphasize the complex nature of the phototropic response, it is quite wrong to assume that a shoot always bends towards the light. It doesn't! What it does depends upon *how much* light it receives. If different samples of oat or corn coleoptiles are given increasing amounts of light, either by increasing the exposure time or by changing the light intensity, they produce different curvature responses. The bending response increases up to a point, but then decreases until at a particular dose level there is no curvature at all. Increasing the light dose still further leads to curvature *away* from the light, then to no curvature, and finally once again to a strong curvature towards the light. This light dose response curve is the most complex known for any photo-biological response, and the reasons for its complexity are not understood. For now we shall consider only the first positive phototropic response to relatively small doses of light: little or nothing is known about the negative and second-position curvatures.

In 1880, Charles Darwin established that the photo-sensing mechanism is located at the extreme apex of the coleoptile, whereas the response takes place in the growing zone some five to ten millimetres lower down. His famous experiments are illustrated here. Tiny tin-foil caps placed over the coleoptile tips to shade the top two millimetres prevented phototropic curvature, even though the growing zone where curvature normally occurred was strongly illuminated from one side. If the two-millimetre apex alone was illuminated from the side, curvature developed in the growing zone, even though it had been kept in the dark. From these experiments Darwin concluded that a message must have been sent from the tip to the growing zone some five millimetres below with instructions to initiate curvature. In fact, Darwin's exact words were, '*we must therefore conclude that some influence is transmitted from the tip to the more basal regions of the shoot*

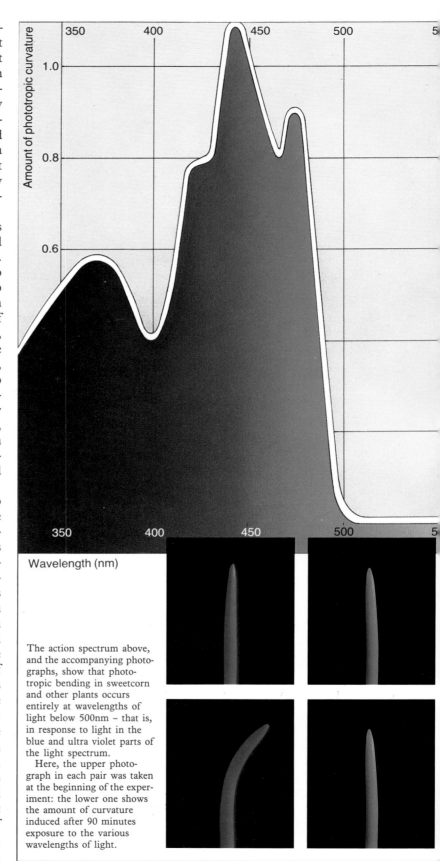

thereby regulating growth and inducing curvature'. This was one of the most significant and profound statements ever made in plant biology, because Darwin's reference to an 'influence' being transmitted was the first recognition that a hormonal type of regulatory mechanism existed in plants – that is, a mechanism in which an active chemical is produced in very small amounts in one part of an organ and then transported to another part where it controls or initiates other processes. We now know that the 'influence' is auxin, or indole-3-acetic acid (IAA).

Having established that the photo-receptor is right at the tip of the coleoptile, we now need to know what wavelengths of light activate it. To answer this question it is necessary to determine an action spectrum – that is, to establish precisely the relative effectiveness, on a quantum basis, of different wavelengths of light. The action spectrum for an oat coleoptile, illustrated here, clearly shows that only blue and near ultra-violet light are active: green, yellow, orange, red and infra-red have no capacity whatsoever to induce phototropic curvature. From these results we can immediately say that the pigment being sought is a yellow, orange or red substance, and this is quite a help since it eliminates pigments like chlorophyll and phytochrome which are green and blue-green substances respectively.

Coleoptiles have been found to contain two chemical compounds that absorb light in the required region of the spectrum. One, riboflavin, is yellow and the other, β-carotene, is orange. The absorption spectra of these substances, however, do not match the oat coleoptile action spectrum in detail. Beta-carotene is a perfect match in the blue region but lacks a peak in the near ultra-violet, while riboflavin has a peak in the near ultra-violet and one in the blue, but the latter lacks completely the detail found in the action spectrum. Years of argument revolved around the relative merits of these two pigments as candidates for the photo-sensing molecule for phototropism. What was largely overlooked however, was the fact that the precise absorption spectrum of a molecule depends, in part, upon the solvent in which it is dissolved. In the cases of the two candidates tested, the riboflavin was dissolved in water, and the β-carotene in an organic solvent! Until recently the natural location of the pigment molecule inside the coleoptile cell was unknown, but experiments focusing on

Wavelength (nm)

The action spectrum above, and the accompanying photographs, show that phototropic bending in sweetcorn and other plants occurs entirely at wavelengths of light below 500nm – that is, in response to light in the blue and ultra violet parts of the light spectrum.

Here, the upper photograph in each pair was taken at the beginning of the experiment: the lower one shows the amount of curvature induced after 90 minutes exposure to the various wavelengths of light.

600

1.0

0.8

0.6

0.4

0.2

0

| 600 | 650 | 700 | 750 | 800 |

this question have now also produced evidence strongly indicating that riboflavin is the active pigment. All pigment molecules that capture light energy to drive or regulate biochemical reactions are highly organized in, or associated with, a membrane. If they were not so organized they would have difficulty in passing on their energy, or signal, to the next stage in the process they regulate. The phototropic receptor pigment is no exception. In recent research, pieces of membrane were isolated from coleoptile cells, and their absorption of light was found to be almost the same as the action spectrum for phototropism. When these membrane fragments were analysed, they were found to contain only a flavin; there were no carotenoids. Thus, evidence from the photoreceptor's natural environment – the cell membrane – strongly points to a flavin as the most likely phototropic pigment. It is now known that photo-activation of this pigment affects the state of oxidation of another molecule, a cytochrome, also located in the membrane. We therefore have a pigment, with the right absorption characteristics, in an organized structure that permits it to pass on its energy to another quite different molecule. This may prove to be the first step in the chain of events which eventually leads to the development of phototropic curvature.

The response mechanism in coleoptiles

The final question concerns the way in which activation of the membrane-bound flavin pigment system brings about the different rates of growth on the two sides of the organ, and hence curvature. In oats and sweetcorn this appears to be achieved by the transportation of auxin from the lighted side of the shoot to the shaded side. Certainly the IAA being transported back from the apex to the growing zone becomes asymmetrically distributed, and there is firm evidence that at least for the first positive phototropic response a lateral transport of this molecule is triggered by the light stimulus. This can be demonstrated by applying small amounts of radioactive IAA to one side of a coleoptile tip and measuring how much is transported across the coleoptile in the dark. The test is then repeated in coleoptiles illuminated from one side. More IAA moves across in the latter case, thus establishing that phototropic stimulation promotes the movement of IAA from the lighted to the shaded side of the coleoptile.

It seems then that the views put forward sixty years ago have been proved correct. Phototropic stimulation, at least for the low-level light stimuli that produce the first positive curvatures, initiates a transport of IAA from the lighted side of the organ to the shaded side. The resulting asymmetry in the distribution of this growth-promoting hormone then produces differential growth which steers the shoot towards the light. But many questions remain unanswered. Is the magnitude of the observed asymmetry in IAA large enough to account for the growth rate changes that occur? Are there other effects such as a change in the rate of longitudinal transport of IAA on the two sides of a phototropically stimulated coleoptile? Could other promoting and inhibiting growth hormones be involved, for it has been found that under some conditions phototropic curvature develops without their being a significant lateral transport of IAA? Finally, no one has yet found the cause of the first negative curvature, or for the second positive curvature.

There is clearly a long way to go before the mechanisms of the photo-sensing guidance systems in plant shoots are fully understood. This is true even in the case of grass and cereal coleoptiles, despite a great deal of concentrated research. We know even less about the photo-sensing mechanisms that operate in the broad-leafed plants!

CHAPTER 8

Red for Go

In the human world a red light is universally recognized as a signal to stop, while its counterpart, a green light, indicates 'go'. In the world of plants, however, things are very different. Here, *red* light is the signal to proceed, whilst green light means little or nothing. It is doubtful whether plants even absorb much green light: the very fact that they appear green means that far from being absorbed, most wavelengths in the green spectral band are reflected straight back from their surfaces.

In nature, plants are exposed only to the 'white' light of the sun, so-called because it consists of a mixture of all the spectral bands or colours. Plants do not actually respond to white light, but rather to particular spectral bands contained in it. In this chapter we shall consider the special role of red light in regulating plant growth and development, but first we need to take a closer look at what happens to the young plant between the time the seed germinates below ground and the moment the plant opens up its foliage to the rays of the sun.

The journey into the light
With the nutrient reserves of the seed mobilized, and growth of the root and shoot under way in the appropriate directions, the next priority is for the shoot to reach the light as quickly as possible. This implies not only the fastest possible rate of growth, but also a very precisely co-ordinated development in which the growing shoot offers as little resistance as possible to soil penetration, and so incurs minimal damage. Penetration of a hostile,

stony, abrasive medium like soil ideally requires a smooth cone-shaped organ, like a root. Shoots, however, are rather different, since even in the embryo they have an apical bud with a number of small foliage leaves already developed, and these have to be carried to the soil surface without damage. If anything as delicate as an embryonic leaf were to project from the apex of a shoot, it would be torn off as the shoot was pushed upwards through the soil. Such a stem would have little or no chance of survival. For safe passage through the soil the shoot must be organized so that development of its leaves and lateral buds is suppressed, while some device is introduced to protect the apical bud and the minute foliage leaves from damage.

Different plants do this in different ways, and the operation of the protective mechanism depends on an environmental signal, namely the presence or absence of light. In darkness, plants adopt a pattern of growth in which the stem grows rapidly and leaf expansion is suppressed. In broad-leaf plants, for example, the apex of the stem is always hooked, so that a perfectly smooth surface is presented for soil penetration. Damage to the delicate apical bud and foliage leaves is eliminated because they are effectively drawn through the soil backwards – through the hole made by the hooked stem. Once the stem apex reaches the light, however, this pattern is completely changed. Stem growth is reduced, the hooked apex straightens out, and the foliage leaves rapidly expand to become effective green photosynthetic organs. So there are two quite different programmes

Germination of the 'Grand Rapids' lettuce seeds opposite could not have occurred without one essential environmental stimulus – exposure of the seeds to red light.

The shoot apex of the French bean (*Phaseolus vulgaris; right*) remains hooked until it breaks through the soil surface and is exposed to the red spectral band in sunlight. This stimulus immediately switches off the dark pattern of growth – the lengthening of the underground stem – and switches on the light pattern in which the leaves expand and the shoot above the cotyledons starts to grow.

78

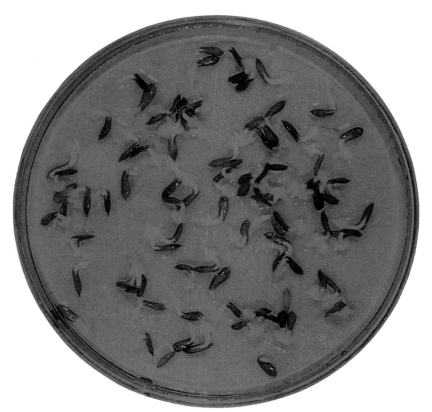

of growth. Which of these programmes is utilized at any particular time depends on whether or not the plant has received a signal: and that signal consists of red light.

Light regulation of growth and development

In the normal course of events, a dicotyledonous seedling will begin to straighten up as soon as its hooked stem breaks the soil surface. However, if seedlings are grown in darkness in the laboratory, they maintain the dark pattern of growth indefinitely – or at least until their food reserves are exhausted. The main characteristics of the dark pattern of growth are the rapid and extensive elongation of the stem, the presence of a hook at the shoot apex, no expansion of the minute foliage leaves already formed around the apical bud, and a characteristic white or slightly yellow colour over the whole shoot due to the absence of the green pigment chlorophyll.

When the uppermost part of the shoot hook breaks through the soil surface, however, it is exposed to light, and this stimulus triggers a totally new developmental pattern. First, the

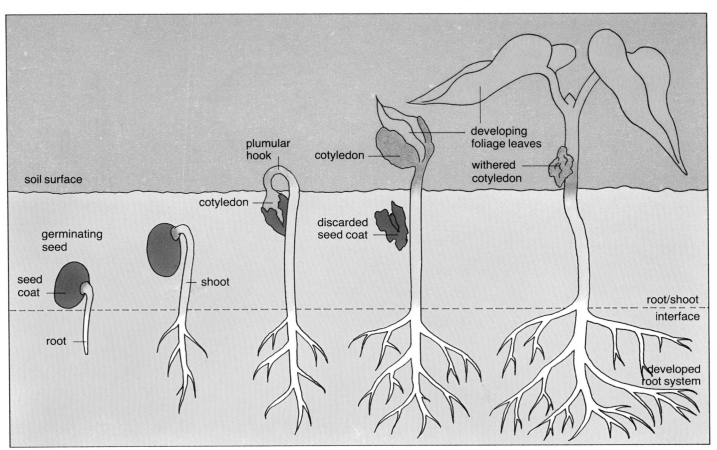

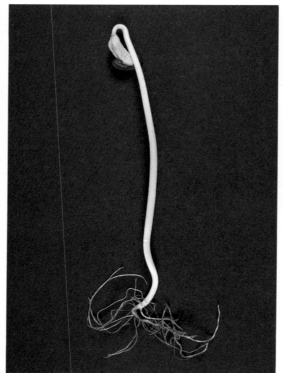

In the light phase of growth (*far left*) a bean shoot's foliage leaves are expanded, the apical bud is upright and the shoot above the cotyledon is growing. In the dark phase (*left* and *above*) only the hypocotyl is growing and the stem apex remains hooked to protect the young leaves, which do not expand.

hooked end of the shoot straightens out so that the apical bud points directly upwards; second, the growth rate of the shoot decreases dramatically; third, the foliage leaves expand rapidly; and lastly, the whole shoot begins to turn green because of the formation of chloroplasts and the synthesis of the green pigment chlorophyll.

While all dicotyledonous and many monocotyledonous plants have these two patterns of growth, grasses and cereals have two rather different patterns, which reflect the different structure of their shoots. The shoot of a grass or cereal seedling consists of a solid stem called a mesocotyl, at the tip of which there is a bud surrounded by small, young, delicate, foliage leaves. The shoot apex is not hooked, and it is certain that there would be extensive, if not fatal, damage to the leaves and apical bud if they were simply pushed up through the soil without some form of protection. The necessary protection for these delicate structures is provided by a curious modification of the outermost leaf at the apex of the stem. This leaf forms itself into a cylindrical structure called a coleoptile, completely surrounding and enclosing the other foliage leaves and the apical bud. The hollow coleoptile is like the finger of a glove, with quite thick side walls, and it has the capacity to grow, and adjust its angle of growth, as necessary to reach the soil surface.

In grasses and cereals the mesocotyl initially grows rapidly beneath the soil, thrusting the coleoptile towards the surface, but as soon as the shoot breaks through, the red light stimulus triggers the change-over to the light pattern of growth.

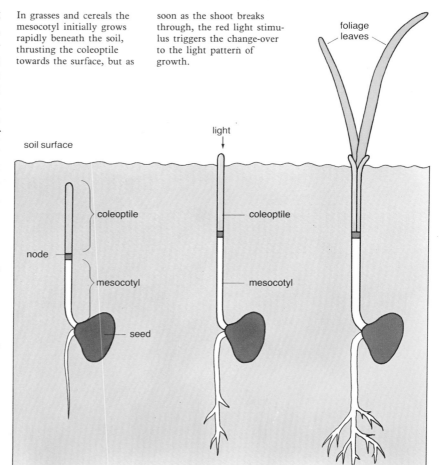

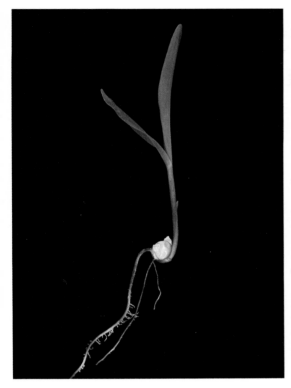

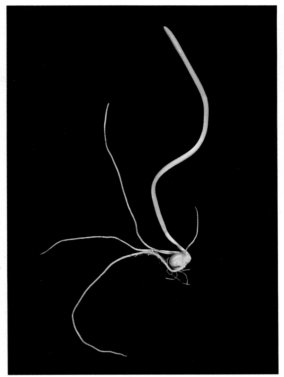

In the light pattern of growth in sweetcorn (*Zea mays; far left*) there is virtually no mesocotyl, the coleoptile is short, and growth is concentrated in the long foliage leaves which have burst through the coleoptile tip. In the dark phase (*left* and *above*) the mesocotyl and coleoptile continue to grow rapidly but the foliage leaves remain small inside the protective coleoptile.

The tip of the coleoptile is rounded and presents a perfectly smooth cone-shaped structure for easy soil penetration. A grass or cereal shoot thus comprises a solid white mesocotyl on top of which sit an apical bud and several foliage leaves enclosed in, and protected by, the tough sheath of the coleoptile.

In the dark, both the mesocotyl and the coleoptile grow rapidly – especially the mesocotyl which pushes the stem apex and foliage leaves towards the soil surface. The young foliage leaves, however, grow more slowly; it is essential that they remain within the coleoptile for their own protection. However, as soon as the tip of the coleoptile penetrates the soil surface, this dark pattern of growth is switched off and a light pattern is switched on. The mesocotyl and coleoptile stop growing immediately and the leaves begin to elongate rapidly. They soon reach the tip of the coleoptile and burst through into the atmosphere where they quickly form chloroplasts, turn green, expand and become functional photosynthetic organs.

The change-over in growth pattern occurs when the shoot tip is first exposed to light on breaking through the soil surface. Those parts of the plant that were beneath the soil remain unchanged: it is the new growth at the shoot apex that follows the new light pattern. If a deeply planted seedling is carefully dug up,

those parts beneath the soil will be elongated and white, while those above will have a normal green appearance.

The dark pattern of growth is known as etiolation, and the phenomenon is widely utilized in gardens and horticultural nurseries. The most succulent rhubarb, for example, is covered over with old chimney pots to 'draw' it, in other words to encourage growth of the leaf stalks in the dark. Similarly, celery is grown in black cylinders to make the stalks white, long and succulent, while asparagus shoots are covered with soil for the same reason.

The one other time at which development will be blocked unless the appropriate light signal is received is when a plant switches over from its vegetative phase to its reproductive phase. A vegetative plant is one that produces only foliage leaves, whereas a reproductive plant produces flowers – which require an entirely different programme of cell development. Not all plants require a special light signal to initiate flower production, but many do, and the nature of the signal is complicated because it really involves the measurement of the time interval between *two* light signals – in other words the measurement of the length of the night. These responses are called photoperiodic responses and they are described in some detail in Chapter 15.

The light-detecting system

Light, as we have seen earlier, can have an effect only if it has been absorbed by a chemical substance. Many substances do not absorb light in the visible part of the spectrum and therefore appear white. Other substances do absorb light, but only in particular spectral bands or colours. Each of these substances will therefore have what is known as a characteristic absorption spectrum, will appear coloured, and will be called a pigment. Chlorophyll, the photosynthetic pigment, for example, absorbs red and blue light and reflects green, hence its green appearance.

The first step in identifying the light-detecting system involved in the control of plant development is therefore to establish precisely which colours (wavelengths) of light produce a response and which do not. A search is then made to find a chemical substance in the plant which absorbs only those 'active' wavelengths.

As we saw in Chapter 5, germination of lettuce seeds is initiated only by wavelengths of light in the orange-red band of the spectrum between 580nm and 700nm, the maximum activity being in the red region at a wavelength of 660nm. Uncurling of the stem apex in dark-grown dicotyledonous seedlings is induced by exactly the same wavelengths, as are the changes in mesocotyl, coleoptile and leaf growth in cereal and grass seedlings. The measurement of night-length in photoperiodic plants involves the measurement of the interval between two red light signals, at dawn and dusk. If the night is interrupted by even a few minutes of light, the pattern of development can be changed, but again it is only red light that is effective. So, we are clearly looking for a pigment that absorbs red light alone, and it should therefore be bluish-green in colour.

Further researches revealed an even more unusual characteristic of this pigment molecule. In studies of lettuce seed germination it was found that the promoting effect of a brief (five minute) exposure to red light could be cancelled if the seeds were exposed immediately afterwards to far-red radiation in the 700–800nm spectral band. So, whatever is switched on by red light to initiate germination can be switched off again by exposure immediately afterwards to far-red radiation. What was even more unexpected was the finding that if a further exposure to red light was given to the lettuce seeds, germination was initiated, while yet another exposure to far-red radiation

again prevented germination. All these exposures to red and far-red radiation must be given immediately one after another, and a number of sequences given to different samples of seeds are illustrated here, together with the results of subsequently leaving the seeds in darkness for two days to ascertain whether or not they would germinate. It is clear that germination can be controlled absolutely by these treatments since germination, or failure to germinate, is determined solely by whether the seeds were last exposed to red or to far-red radiation.

The same situation exists with all the other developmental responses in which the red signal is involved, such as the straightening of the hooked stem and the expansion of the leaves. What is important is that red light initiates the changes, far-red radiation can nullify the effect of red light, and, most importantly, far-red radiation alone does not affect a dark-grown plant in any way.

These experimental findings enable us to make a number of important deductions about the characteristics of the light-detecting system. First, there must be a pigment in plants which absorbs red light. Second, the pigment molecule can evidently exist in two freely-interchangeable forms, one of which absorbs red light and the other far-red radiation: the two forms must therefore have slightly different chemical structures. Third, the active form of the pigment, which initiates the changes in growth and development, must be the far-red absorbing form.

This pigment is called phytochrome, and its two forms are usually represented as P_{red} and $P_{far-red}$ (or P_{660} and P_{730} after the wavelengths of radiation the two forms absorb most effectively). The interconversion of the two forms is usually summarized as follows:

$$P_{red} \underset{\text{Far-red radiation}}{\overset{\text{Red light}}{\rightleftharpoons}} P_{far-red} \rightarrow \text{Development changes}$$

After exposure to red light the pigment will be mostly in the far-red absorbing form ($P_{far-red}$) and changes in development will be initiated in the plant. After exposure to far-red radiation the pigment will be in the red absorbing form (P_{red}) and no changes will be initiated; the plant behaves as though it is in undisturbed darkness. Because seeds and plants that have never been

dark

red

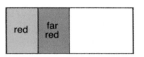
red | far red

red | far red | red

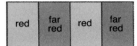
red | far red | red | far red

The complete photo-control of germination in 'Grand Rapids' lettuce seeds is demonstrated by exposing the seeds sequentially to red and far-red radiation and checking how many of the seeds germinate.

In total darkness only 1% to 2% germinate. Brief exposure to red light causes about 95% germination. Exposure to red and then far-red radiation again causes only 1% to 2% to germinate.

Any number of such alternating five-minute exposures can be made: if the last one is red, the seeds will germinate, if it is far-red they will not.

exposed to light respond only to red light and not to far-red radiation, the pigment in such plants must all be in the red absorbing form. Seeds with a light requirement for germination become sensitive to red light only after they have taken up water.

In nature, of course, neither seeds nor seedlings are exposed separately to red and far-red radiation; these two spectral bands occur in about equal amounts in sunlight. So a seed which has imbibed water on, and a growing shoot breaking through, the surface of the soil will be exposed simultaneously to equal amounts of red and far-red radiation. Since germination of light-requiring seeds and other plant developmental changes are initiated by sunlight, the reader may wonder why the effectiveness of the red spectral band in sunlight is not abolished by the far-red spectral band. It is because in a newly imbibed seed or seedling that has never before been exposed to light the pigment is all in the red-absorbing form: the plants therefore have the capacity of absorbing and detecting only the red spectral band in the sunlight – they can not detect, or 'see', the far-red band because they have no means of absorbing it. At first, then, sunlight acts essentially as pure red light. After a few seconds of exposure to sunlight, however, those molecules of the red-absorbing form of the pigment which have absorbed light in the red spectral band will have been converted to the far-red absorbing form of the pigment, and the plant will have acquired, for the first time, the capacity to detect the far-red spectral band in the sunlight. As these ($P_{far-red}$) molecules begin to absorb the far-red spectral band they will be converted back to the red absorbing form of the pigment again. The rates of the forward and backward conversion reactions eventually reach steady values and the pigment system comes to what is known as a photo-stationary

steady state, in which the total amount of the pigment in each form is constant, even though the molecules are being converted backwards and forwards all the time. What is important about the photo-stationary state is that a substantial proportion of the pigment is in the far-red absorbing form, thus initiating germination and developmental changes.

One other important physiological fact must be explained before we can understand how the phytochrome system works. So far it has always been carefully stressed in discussing the reversible effects of exposures to red and far-red radiation that the exposures should occur immediately one after another. If this is the case, germination and other developmental patterns remain totally under the control of the phytochrome pigment. But what happens, for example, if lettuce seeds are given a five-minute exposure to red light and are then left in darkness for half an hour before being given a five-minute exposure to far-red radiation? The answer is that the half-hour interval of darkness renders the far-red treatment totally ineffective, and the seeds germinate. The inference is that if the far-red absorbing form of the pigment is made, and left in the plant for about 30 minutes, then converting the pigment back to the red-absorbing form no longer prevents germination from taking place. In other words, the germination process has escaped from the control of the phytochrome pigment. The same effect is seen in the developmental changes that occur when a shoot is first exposed to light, but some other processes, like photoperiodic responses, seem to remain permanently under phytochrome control.

Escape from photocontrol can best be illustrated by considering horses in a stable. When the door is shut, the horses can not get out. If the door is opened for a few minutes, it is unlikely that any will escape, and the door can be closed again with all the horses still inside. However, if the door is opened, and left open for 30 minutes, when you return you can close and open the door as much as you like but it will no longer have any control over the horses because they will all have escaped into the field! The open door can be equated to the far-red absorbing form of phytochrome, the closed door to the red absorbing form and the escape of the horses to the process of germination.

In more scientific terms it means that the far-red absorbing form of the pigment probably promotes the synthesis or release of chemical substances that initiate other reactions, such as those leading to germination, and that once enough of these chemical substances are present in the plant cells to initiate these other reactions, the whole process no longer depends on the presence of the far-red absorbing form of the pigment. Converting it back to the red absorbing form therefore has no effect.

All these experiments indicate the presence in plants of a unique pigment which can exist in two interchangeable forms, one absorbing red and the other far-red radiation. All we have to do now is grind up, and analyse, a selection of plant tissues and establish whether such a pigment really exists – and if so, identify it.

Identification of phytochrome
Since phytochrome absorbs only red and far-red radiation it will reflect blue and green light and should therefore be bluish-green in colour. If plants are ground up into a slurry and then processed to separate out the different chemical substances present, it is found that a fraction which contains soluble proteins takes on a bluish-green colour as it is concentrated. Could this fraction contain phytochrome? The way to establish whether its bluish-green colour is due to the presence of the pigment that underlies the photo-responses described earlier is to ascertain whether or not it absorbs light in precisely those spectral bands that produced the developmental changes in the plants.

So first we need to know exactly what the action spectrum looks like for the developmental response in the plant. Action spectra were discussed in Chapter 7 in the context of the plant's light-sensing guidance system, and in that case it was shown that only the blue and ultra-violet bands of the spectrum were effective. The action spectrum for light-induced germination of lettuce seeds was shown at the end of Chapter 5, and action spectra for the

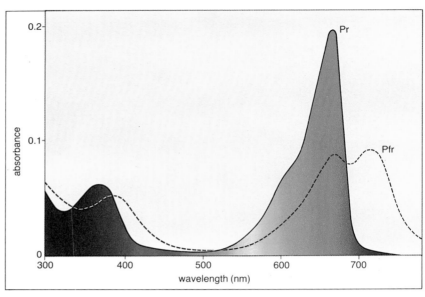

other developmental responses mentioned in this chapter are virtually identical. It is immediately clear that activity is wholly confined to the spectral band from 580nm to 700nm, with a peak of activity at 660nm. We are therefore looking for a pigment molecule that absorbs light only between 580nm and 700nm and which has a maximum absorption at 660nm; in other words a pigment having an absorption spectrum identical to the action spectrum. The bluish-green substance extracted from seedlings grown in total darkness has just such an absorption spectrum; strong evidence indeed that we have phytochrome in the test tube.

But phytochrome must be able to change its absorption spectrum from one having a peak absorption in the red band to one having a peak absorption in the far-red band. We therefore have to establish that the absorption spectrum of the extracted pigment changes in this way when it is irradiated with red light.

The action spectrum for the reversal of the effect of red light by a second exposure to radiation of different wavelengths shows a peak of activity at 730nm – in the far-red band. If the extracted pigment in the test tube, which was clearly the red-absorbing form, is now exposed to a strong red light for 15 minutes and its absorption spectrum re-examined, it is found to have changed. Peak absorption has moved from 660nm to 730nm, and the pigment is now obviously in the far-red absorbing form. The solution in the test tube is also found to have changed colour slightly, as would be expected with a change in the absorption characteristics of the chemical molecule.

The absorption spectrum of phytochrome obtained from oat seedlings when the chemical is in the red-absorbing form (solid line) and in the far-red absorbing form (dotted line). It is clear that the red-absorbing form will not absorb radiation in the far-red region of the spectrum – that is, above 700nm.

The test tubes below contain solutions of the light-absorbing pigment phytochrome in its two interchangeable forms – the bluish-coloured red-absorbing form on the left and the greenish-coloured far-red form on the right.

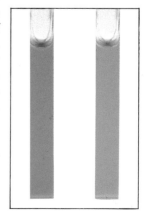

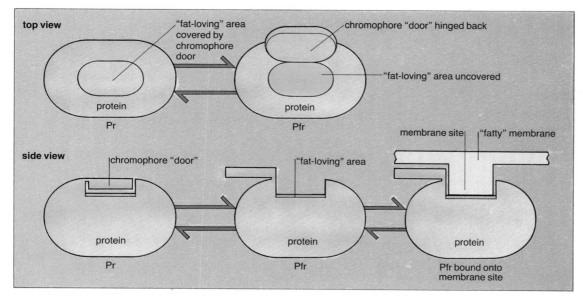

top view

"fat-loving" area covered by chromophore door

protein

Pr

chromophore "door" hinged back

"fat-loving" area uncovered

protein

Pfr

side view

chromophore "door"

protein

Pr

"fat-loving" area

protein

Pfr

membrane site | "fatty" membrane

protein

Pfr bound onto membrane site

One theory of how the phytochrome molecule works (*left*) suggests that in the red-absorbing form, part of the molecule called the chromophore covers up a fat-loving 'sticky patch' and that in the far-red-absorbing form, this chromophore lid swings back to expose the fat-loving area which can then bind on to an appropriate site on the fatty cell membrane.

Finally, if the far-red absorbing form of the pigment in the test tube is now exposed to strong far-red radiation for 15 to 20 minutes, and its absorption spectrum re-examined, it is found to have reverted to the red-absorbing form with a peak at 660nm.

There is now no doubt that our test tube contains phytochrome. But what, precisely, *is* phytochrome? Detailed analysis has shown it to be a large molecule consisting of two distinct parts. The larger portion consists of protein while the smaller portion – called a chromophore – is a special kind of molecule found in a number of plant pigments. The key to the operation of the phytochrome system seems to be that conversion of the pigment from the red to the far-red absorbing form causes the chromophore to change shape so that instead of floating freely in the cell cytoplasm the whole phytochrome molecule binds on to the membranes of the various structures within the cell, dramatically altering their electrical properties and their permeability. The electrical charge across the plasmalemma, for example, can change within 15 seconds of the cell being exposed to red light!

Other phytochrome responses, such as seed germination and the straightening of the dicotyledonous shoot's hooked stem, are much slower – often developing gradually some hours after the red light stimulus has been received. These slower responses are thought to be caused by the active form of the pigment initiating a leakage, from some of the subcellular particles, of special chemicals which in turn alter the plant's growth pattern. It has been shown, for example, that plastids in the leaves of wheat seedlings grown in darkness contain a high level of gibberellins, but that these molecules do not leak out through the membrane surrounding the plastids if they are kept in darkness. However, after a brief exposure to red light, these powerful growth-regulating hormones leak out through the membranes quite readily.

The far-red form of phytochrome also combines with other molecules in the plant to produce active compounds capable of switching on changes in cell chemistry. More than 50 different enzymes are now known to be under phytochrome control – the manufacture of some being initiated, that of others halted, simply by exposure to red light. There is now strong evidence that phytochrome controls the manufacture of enzymes by regulating access to the genetic information stored on the chromosomes in the cell nucleus. It appears to do this by controlling the formation of messenger RNA, used by the ribosomes as their blueprint for enzyme synthesis. By controlling the process at this critical 'transcription' stage, the phytochrome regulates which pieces of genetic information are available at any particular time – and hence which developmental programme the plant will follow.

Much more research will be necessary before we achieve a full understanding of the complexities of the phytochrome pigment system. One thing, however, is beyond doubt – and that is that red light has a profound and fundamental influence on many of the biochemical processes that take place in plant cells.

These photographs illustrate the kind of thing that happens in the Tanada effect which demonstrates the dramatic changes that occur in the electrical charge on root surfaces when phytochrome is converted from the red to far-red absorbing forms and vice versa. Root tips of barley and mung bean stick to negatively charged glass surfaces after exposure to red light (*top*) but not after exposure to far-red.

CHAPTER 9

Photosynthesis – The Capture of Solar Energy

Photosynthesis is the process by which plants capture and package the energy in the sun's rays. It can then be transported around the plant, to be released when and where it is required to do useful work, like driving chemical reactions. In practice, this involves storing the energy in chemical molecules that are stable, but that can, in the presence of appropriate enzymes, be broken down to release the energy they contain. It is also the means whereby the plant acquires, from the air, the carbon atoms that are the principal components of the molecules that make up its body.

Photosynthesis can be summarized as a chemical reaction in which carbon dioxide from the air, and water, react to produce carbohydrate (sugar and starch) and oxygen according to the general equation:

$$CO_2 + H_2O \longrightarrow [CH_2O] + O_2$$

This equation is, however, rather misleading, because carbon dioxide and water do not react when simply mixed together! There must, therefore, be a great deal more involved. Photosynthesis takes place only in the green parts of plants, in light, so the green pigment chlorophyll, and light, are clearly both necessary. But even if the equation above does describe the overall effect of photosynthesis, an interesting question should immediately spring to mind, and that concerns the source of the oxygen that is evolved. Both carbon dioxide and water contain oxygen, so from which molecule does the evolved oxygen gas arise? The importance of this question is that the answer will tell us whether photosynthesis involves splitting the

water molecule and adding the hydrogen atoms to the carbon dioxide molecule, or alternatively splitting the CO_2 molecule and adding the carbon to the water molecule. Already we are asking quite deep questions about the mechanism of photosynthesis – but we are moving a little too quickly. Before delving into the process itself we must first examine the structure of the plant organ in which this life-sustaining process principally takes place.

The leaf

In most plants photosynthesis takes place in the leaves, although in some, like the cacti which have dispensed with leaves because they are too vulnerable to desiccation, photosynthesis is carried on wholly in the stem. In these plants, the stem is often flattened so that it looks slightly leaf-like and presents a large surface to the sun's rays. In some epiphytic plants, such as the orchids that grow high in the branches of tropical forest trees, even the long trailing aerial roots turn green and carry out photosynthesis. Flower parts, too, can carry out photosynthesis, perhaps the most familiar examples being the green, photosynthetically active sepals and the 'pods' of peas and beans.

In most plants, however, the leaf is the main site of photosynthesis, and the typical leaf is very highly specialized for this task. It has a large surface area with which to intercept the sun's rays, and many plants have mechanisms with which they can adjust the leaf's position so that its surface is always held at right-angles to the incident light. The outer surface of the leaf is perforated with a large number of minute

Foliage, sunlight, water and air – all the requirements of photosynthesis portrayed in a single photograph taken in the rainforest of Costa Rica.

Photosynthesis in organ-pipe cacti (*far left*) is carried out entirely in the green stems. Leaves are far too delicate and vulnerable to desiccation in the hot, dry, air of the desert environment.

Even the roots are able to carry out photosynthesis in epiphytic plants such as the orchid *Taeniophyllum* (*left*). This species does not produce any photosynthetic leaves at all.

Under a microscope (*below* and *below left*) a section of a privet leaf reveals the vascular transport tissues of the veins, and the palisade and mesophyll tissues of the leaf blade. Note the difference between upper and lower epidermis.

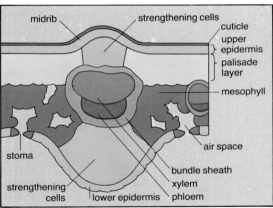

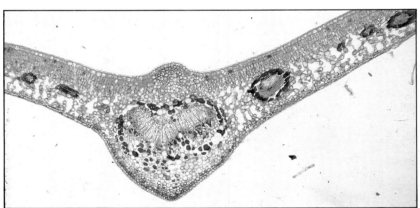

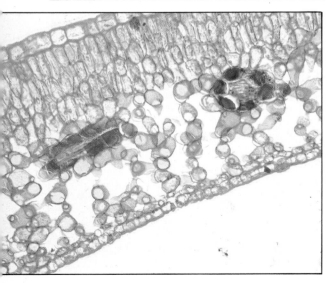

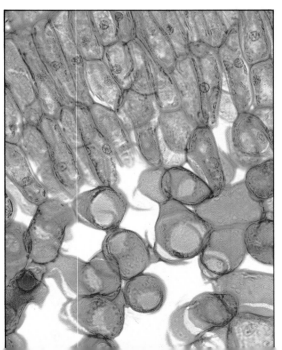

At increasingly higher magnifications (*far left* and *left*) the tall, thin columnar palisade cells and their chloroplasts are clearly visible in the upper half of the leaf. The lower half of the leaf consists of mesophyll cells, separated by large, irregular air spaces.

pores called stomata, which can open and close. When open they allow inward movement of the CO_2 necessary for photosynthesis; when closed they limit water loss from the leaf – a problem which all plants have to deal with in daytime. Leaves are also well supplied with veins connected to the main 'plumbing system' of the stem and roots (Chapter 10). This ensures that they have a plentiful supply of water drawn up from the soil, and also an efficient method of carrying the products of photosynthesis from the leaves to other parts of the plant, where they may be utilized at once or stored in the form of starch.

The outer surface of a typical leaf is covered with a waxy layer, the cuticle, the function of which is to make the surface impermeable to water and so protect the plant from excessive water loss. In addition, the leaves of some plants have hairs on the surface. These reduce the flow of air over the leaf, creating a thicker, unstirred layer of air which reduces the rate at which water is lost through the stomatal pores.

Surface hairs may also have a protective function – as contact with those of a stinging nettle will quickly demonstrate!

The outermost layer of cells in a leaf – the epidermis – is a relatively complicated but important layer consisting primarily of quite ordinary cells which, in many plants, look in surface view like crazy paving. They are tightly and strongly joined together and they do not have chloroplasts. Distributed amongst these epidermal cells are the stomatal pores which allow the movement of gases into and out of the leaf. Each pore is flanked by, indeed is formed by, two highly specialized, sausage-shaped guard cells, well supplied with chloroplasts. Because of the uneven thickening of their cell walls, these guard cells change their shape according to environmental conditions. When they are pumped up to full internal hydraulic pressure they become banana-shaped, with the result that their adjacent walls are separated – thus creating an open pore in the epidermal layer. On losing their internal pressure, the cells straighten out and their walls come together again, so closing the pore. It is solely these changes in the hydraulic pressure in the guard cells that determine whether the pores are open or closed. In dry weather, water is lost from the plant quite rapidly, and the guard cells tend to lose pressure quite early: the result is that the pores close, and water loss by the plant is drastically reduced before serious damage occurs.

Inside a dicotyledonous leaf there are usually two contrasting regions of tissue. In the upper half there is a tightly packed layer of long thin cells called the palisade layer. The cells of this

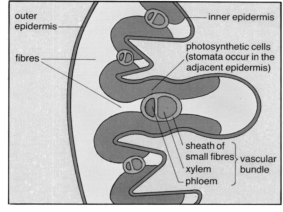

A magnified section across a grass leaf reveals the girder-like structure of the supporting tissues (red) extending from the vascular bundles to the epidermis (*right* and *below*). Also visible is the green photosynthetic tissue, which is often protected from drying by the curling of the leaf. To reduce water loss, stomata occur only in the inner grooves.

outer epidermis — inner epidermis

fibres

photosynthetic cells (stomata occur in the adjacent epidermis)

sheath of small fibres
xylem } vascular bundle
phloem

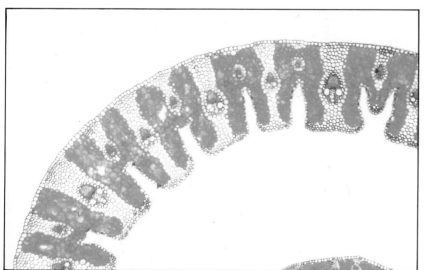

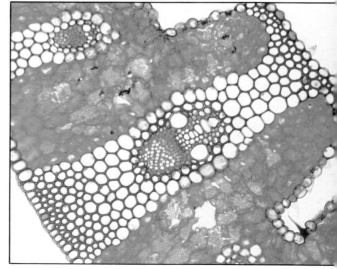

This close-up shows one of the stomatal pores on the underside of a *Rhoeo* leaf. The pore is formed by the sausage-shaped guard cells which, by adjusting their internal hydraulic pressure, make the pore open and close. The guard cells contain numerous chloroplasts but the surrounding subsidiary cells are like other epidermal cells in having no chloroplasts.

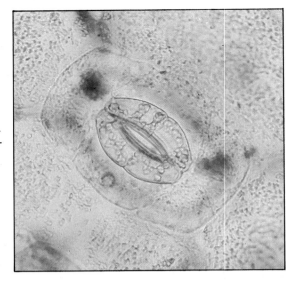

The *Narcissus* has stomata in the epidermis of both upper and lower surfaces of its leaves. Note the vascular bundle and the palisade cells beneath the epidermis of both surfaces.

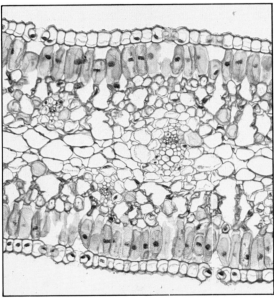

Higher magnification of an individual pore reveals that the guard cells have tiny transparent projections at the inner and outer ends of the pore. These anti-vortex devices prevent moisture being drawn out of the leaf in high winds.

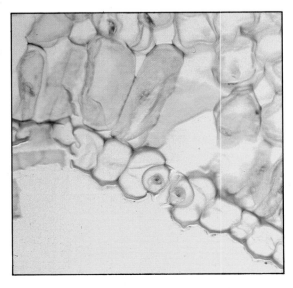

layer contain many chloroplasts, and they are the principal site of photosynthesis. The lower half, the mesophyll, comprises a large number of very irregular cells with large air spaces between them. Threading their way amongst these cells are the veins containing the transport tissues xylem and phloem, the former carrying the leaf's water supply to maintain the hydraulic pressure of the cells and fuel the process of photosynthesis; the latter transporting away the leaf's photosynthetic products.

The leaves of grass and cereal plants are somewhat similar to those of dicotyledonous plants except that their internal cells are not differentiated into palisade and mesophyll tissues and there tends to be much more strengthening tissue along and around the veins. Many monocotyledonous leaves are very long and narrow, as in the various species of iris, and the grasses, so such strengthening tissue is of considerable importance. A very important difference between dicotyledonous leaves and those of grasses and cereals is that in the latter the growing zone is located at the *base* of the organ whereas in the former it is not so localized. This is why grasses can be grazed by domestic and wild animals without suffering permanent damage. The animals can feed without removing the vital growing zone of the plant. By contrast, dicotyledonous plants are for the most part seriously damaged or killed by grazing animals unless they have become highly specialized, for example by keeping their leaves pressed flat against the surface of the soil, as in the case of the daisy, or by having enormous power to regenerate new leaves from an underground stem or root stock.

Photosynthesis – the light and dark reactions

Soon after the start of the present century, scientists discovered that photosynthesis did not just consist of a single light reaction, but included at least one reaction that went on in the dark. To appreciate how this was deduced, we need to know two facts of physical chemistry. Firstly, a photochemical, or 'light', reaction depends *only* upon light and is not influenced by temperature. Therefore, at a particular light intensity the rate of the reaction will be the same at widely different temperatures and will change only if the light intensity is altered. Secondly, a non-photochemical, or 'dark', chemical reaction *is* affected by temperature, and the rate of the reaction will double or in some cases triple for an increase in temperature of 10°C.

Studies of the rate of photosynthesis in plants at different light intensities and temperatures revealed that at a constant low light intensity, increasing or decreasing the temperature had no effect on the rate of photosynthesis. This indicated that the rate was being determined by the low light intensity which was limiting the rate of a photochemical or light reaction. At a constant high light intensity, however, the rate of photosynthesis was increased if the temperature was increased, indicating that the process was being limited not by the light intensity but by the rate of a dark chemical reaction which was sensitive to temperature.

More sophisticated experiments some thirty years later utilized flashing light. It was found that the amount of photosynthesis per flash depended on the intensity of the flash and also on the interval of darkness between the flashes. Maximum photosynthesis was achieved when the dark interval was increased to about 0.1 seconds, clear evidence that it took this amount of time for the dark reactions to proceed to completion following a very short flash of light. It is clear, then, that there are two parts to the photosynthetic process – the light reactions and the dark reactions.

Before we examine the process in detail, it might help the reader to visualize the mechanism of photosynthesis as a two-stage process in which light reactions first capture the energy of the sun, package it up neatly, and place the packages on a conveyor belt for transfer to another part of the chloroplast. On arrival at their destination the packages are opened, and their energy is released and repackaged. This repackaging is achieved by a series of dark chemical reactions driven by some of the released energy. The dark reactions lead to the formation of sugar and starch, and it is in these stable compounds that the energy is stored for future use.

The light reactions

In all plants except the red algae photosynthesis takes place exclusively in the sub-cellular particles known as chloroplasts. These are complex structures which have a very intricate arrangement of internal membranes upon which photosynthesis depends.

In a young seedling leaf that has not been exposed to light, the cells contain no recognizable chloroplasts and no chlorophyll. Instead there are bodies called etioplasts, or proplastids, which are surrounded by a double membrane and contain a large crystalline structure called a prolamellar body. On exposure to light this prolamellar body disappears and gives rise to a large amount of membrane material which becomes organized in a very special way, and

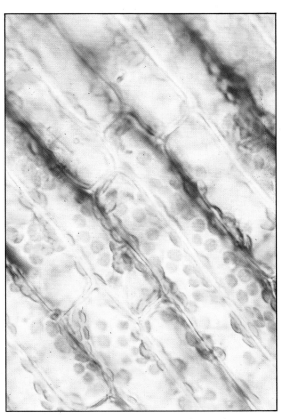

In this magnified surface view of an *Elodea* leaf, the chloroplasts are clearly visible inside the cells.

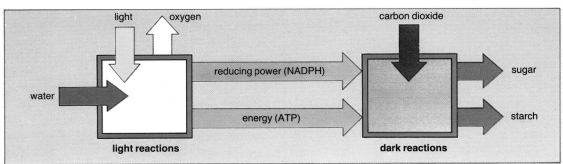

The two main parts of the photosynthetic process are the light reactions (left box) and the dark reactions (right box). In the left box, solar energy is captured, water is split, and oxygen is released. The stored energy (NADPH and ATP) is transferred to the dark box where it is used to capture carbon dioxide for the production of sugar, starch and other carbohydrates.

91

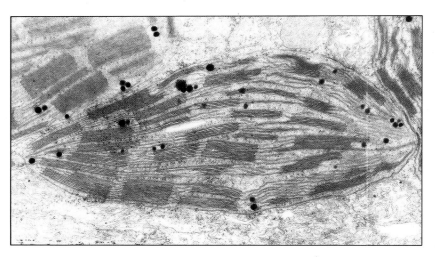

The power of the electron microscope reveals the detailed inner structure of a chloroplast. Inside the double outer membrane are the stacks of dark, flattened thylakoid vesicles that form the grana. These are linked by the lighter-coloured membranes that pass through the stroma.

in which chlorophyll molecules become deposited to form the normal green chloroplast.

A mature chloroplast is surrounded by a double membrane similar in composition to the plasmalemma and other cell membranes. The internal membranes become organized into flattened vesicles, like squashed balloons, piled up into stacks. The individual vesicles are called thylakoids and it is in the thylakoid membranes that chlorophyll and the other photosynthetic pigments are located. The stacks of rather uniform thylakoids form what are known as grana, and these are embedded in the semi-liquid contents of the chloroplast called the stroma. Some rather larger membranous sacs run from one granum to another, and these are called stroma thylakoids. The light reactions in photosynthesis all occur in the thylakoid membranes whereas the dark reactions all take place in the stroma.

Capturing the sun's energy

All radiation consists of minute packages of energy called quanta, or in the case of light energy, photons, and in order for the sun's radiant energy to be captured, these photons must be absorbed by chemical molecules. The so-called 'white' light of the sun is in fact a mixture of all the different colours of light, each colour being due to radiation of a different wavelength. Blue light, for example, has wavelengths in the range 400nm to 500nm, while red light ranges from 600nm to 700nm. Now, in addition to having different colours, the different wavelengths of light also differ in the amount of energy contained in their photons. A single photon of ultra-violet light, for example, contains about twice as much energy as a photon of red light – which is why ultra-

violet light can be so damaging to living tissues. High-energy photons can cause molecules to disintegrate if they are absorbed, and most organic molecules absorb ultra violet light! Lower-energy photons can be absorbed without damage, but the energy gained by the molecule leads to it being converted to what is known as the excited state. This conversion is achieved by an electron in the outermost part of the molecule being pushed out still farther, into a new orbit. Such a precise adjustment in the electron orbit requires a precise amount of energy, and this is why only certain photons are absorbed by particular molecules. Photons with too much or too little energy cannot make the required adjustment in the electron orbit and so are not absorbed; those with just the right amount of energy are absorbed. For this reason many molecules absorb certain colours of light and not others, and such molecules are termed pigments.

When a pigment molecule absorbs a photon of light and is transformed into the excited state, it contains more energy than it did when it was in the ground state. In the case of chlorophyll, the main photosynthetic pigment, there are in fact a number of excited states, but we need only be concerned with one – known technically as the first excited singlet state. Pigment molecules raised to this state exist there for only a very short time. They quickly lose their newly-acquired extra energy and return to the ground state – after a 'life' of about a one-thousand-millionth of a second! But it is *how* their extra energy is lost that is important, for if it can be passed on to another chemical substance then it opens up the possibility of the plant retaining the energy so recently and temporarily captured by the chlorophyll molecule.

There are a number of ways in which an excited chlorophyll molecule can lose its energy and return to the ground state. The first is by simply losing its energy as heat; the second by re-emitting the energy as a photon of light in the process known as fluorescence; the third by transferring the energy to another neighbouring molecule; and the fourth by becoming involved in a chemical reaction. It is the last two kinds of energy loss that are of critical importance in photosynthesis, though the other two occur to a limited extent as well. We shall discuss these questions of energy transfer later, after we have discovered a little more about the pigments involved in photosynthesis.

The photosynthetic pigments

The wavelengths of light most active in bringing about photosynthesis are those in the red and blue spectral bands. The pigment molecules that capture the sun's energy will therefore have to show peaks of absorption in these bands, and will thus be green in colour. Extraction of the green pigments from higher plants has revealed that there are two types of chlorophyll, called chlorophyll-a and chlorophyll-b, which differ slightly in their absorption spectra. There are, however, other yellow-orange pigments called carotenoids which also absorb light and participate in driving photosynthesis.

The chlorophyll molecule consists of two distinct parts, a 'head' consisting of a ring-shaped chemical structure with a magnesium ion at its centre, and a long straight 'tail' consisting of a phytol chain. The head is responsible for light absorption and the 'tail' appears to anchor the molecule in the thylakoid membrane. Chlorophylls a and b differ only in the nature of a small chemical group attached to one part of the head; a $-CH_3$ group in chlorophyll-a being replaced by a $-CHO$ group in chlorophyll-b. The carotenoids are quite different, linear, molecules with a small ring at each end. These molecules absorb only in the blue spectral band, as the reader may recall from Chapter 7 where we discussed their possible involvement in phototropism.

These photosynthetic pigments are embedded in the thylakoid membranes. The hydrophobic (water hating) phytol tail of the chlorophyll molecule has a particular affinity for the lipid membrane while the hydrophilic (water loving) head remains on the surface. The whole of the carotenoid molecule is hydrophobic and is therefore probably completely embedded in the lipid membrane. The orientation of these molecules in the thylakoid membrane, and their close association with neighbouring molecules, is of critical importance in the capture, transfer and storage of the

The action spectrum on the right shows that the process of photosynthesis is driven principally by blue and red light.

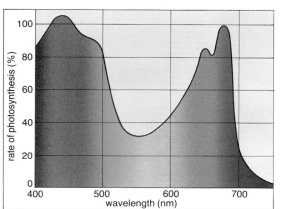

The main pigments involved in capturing the sun's energy to drive the photosynthetic process must absorb light of red and blue wavelengths. Here (*right*) the absorption spectra of chlorophyll-a (blue line) and β-carotene (red line) show that they fit the requirement exactly.

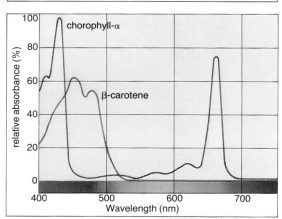

The molecular structures of chlorophyll-a and β-carotene are illustrated (*far right*) to show the complexity of these light-absorbing pigments. The stylized diagram below shows how they are thought to be arranged in the thylakoid membrane.

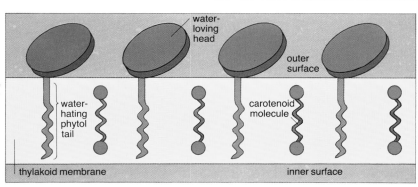

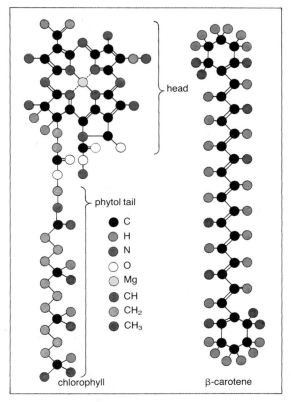

sun's energy. But these molecules do not all participate directly in driving the process of photosynthesis. Most of them are essentially just light traps, and having captured a photon of energy, they immediately pass it on from one molecule to another until it reaches a special chlorophyll molecule which functions as a reaction centre. It is here that the first real reaction of photosynthesis occurs. The collecting pigment molecules and the reaction centre molecule together make up a highly organized structure called a photosynthetic unit.

The photosynthetic unit

A photosynthetic unit can be likened to the dish aerials used to collect television signals or, on a much larger scale, to collect radio signals from space. The dish collects the signals and concentrates them on to a sensor which then converts them into electrical impulses to be fed into a television set or computer. On the thylakoid membrane, each 'dish' consists of about 300 molecules of chlorophyll-a and chlorophyll-b, together with some carotenoids. Together, these pigments are called the antennae pigments. They are arranged around a special chlorophyll-a molecule which alone appears to have the capacity to process the captured light energy and bring about the first reaction of photosynthesis. The antennae pigments effectively form a tuned aerial which captures solar photons of particular energy levels. The captured energy then passes from one molecule to another, without energy loss, until it reaches the reaction centre.

The situation is, however, complicated by the fact that not one photochemical reaction, but two, are involved in the process of photosynthesis, and it may help our understanding if we liken them to a television set that requires two slightly different dish aerials to make it work. The first dish passes the signal it receives to the second dish where it is enhanced by the signal captured by the second dish before being passed on to the electronics of the television set. In photosynthesis, the two photosynthetic units, the equivalent of the two dish aerials, are called Photosystem I and Photosystem II, and each consists of an array of antennae pigments and a reaction centre. The antennae pigments are virtually identical in the two photosynthetic units, but the special chlorophyll-a molecules in the reaction centres are slightly different.

The two photosystems and their interaction

The action spectrum for photosynthesis has peaks in both the blue and the red spectral bands and its shape is not identical to that of the absorption spectra of any of the principal photosynthetic pigments. This fact alone indicates the possibility of more than one pigment being involved. The action spectrum also shows a very important feature in the red region – a sharp decline in photosynthesis as the wavelength of light is increased from 680nm to 720nm, despite these wavelengths being absorbed quite strongly by chlorophyll. It was the study of this 'red drop' effect that provided the first conclusive evidence that two light reactions were involved in photosynthesis. If a plant was exposed to a beam of the low efficiency red light at, let us say, 700nm, and then simultaneously exposed to another beam of shorter wavelength red light at, for example, 650nm, the photosynthetic rate increased to the level that would be expected on the basis of the intensity of the 700nm radiation. In fact, its value exceeded the sum of the values given by the 650nm and 700nm spectral bands separately. This phenomenon, the Emerson enhancement effect, shows conclusively that there are two light-driven processes in photosynthesis. In higher plants these two light reactions are carried out in the two distinct photosynthetic units called Photosystem I and Photosystem II. The chlorophyll-a molecules in the reaction centres of these two photosystems are slightly different, and show peak absorption in the red spectral band at slightly different wavelengths. For this reason the chlorophyll-a molecule at the reaction centre of Photosystem I is called P700 and that at the reaction centre of Photosystem II is called P680, the number indicating the wavelength of maximum absorption. The Emerson enhancement effect is thus explained by the fact that at wavelengths above about 690nm Photosystem II does not operate, and photosynthesis becomes very inefficient because Photosystem I, which absorbs up to about 720nm, cannot really function alone. The second beam of red light at 650nm has the effect of bringing Photosystem II into operation, thus enabling photosynthesis to take place because *both* photosystems are able to capture light energy. We must now consider how these two types of photosynthetic units co-operate, and in doing so we shall discover the source of the oxygen that is liberated during photosynthesis.

This diagram illustrates the *two* photochemical reactions in photosynthesis and the way in which they are linked together. The overall effect is to split water into its component oxygen and hydrogen, then release the oxygen as gas and store the energy captured from the sun in the form of reducing power (NADPH) and a concentration gradient of hydrogen ions across the thylakoid membrane.

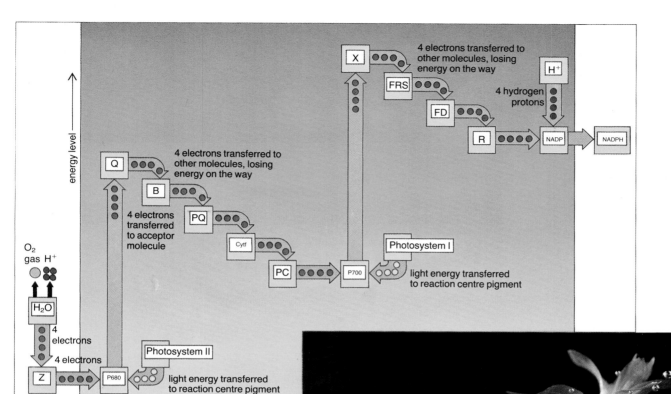

When a photon of light energy is captured by the antennae pigments of Photosystem II, it is transferred to the reaction centre pigment P680. As a result, this pigment is converted to its excited state, and an electron is pushed out into an orbit more distant from the centre of the molecule. In this excited state the electron is actually lost altogether to another acceptor molecule, called Q, and the P680 molecule effectively steals an electron from a nearby molecule called Z to replace the one it has lost. But having collected this electron, the P680 molecule is now once more in the unexcited ground state. It can therefore absorb another photon of light, lose another electron, and once again 'steal' a replacement electron from Z. By the time Z has lost four electrons it has developed four positive charges and in this state it can actually withdraw four electrons from a molecule of water – effectively splitting the molecule into oxygen, which escapes as a gas, and positively charged hydrogen ions (H^+) which remain in solution inside the thylakoid vesicle. So now we know that the oxygen evolved in photosynthesis comes from water and *not* from carbon dioxide.

Returning briefly to the electrons that were lost by the P680 pigment and gained by the molecule labelled Q; these electrons do not remain in Q but begin to tumble 'downhill' by being transferred, in succession, to at least four other molecules, losing some of their energy in the course of each transfer. The chemical identities of these substances are well known, but for simplicity they are designated B, PQ, Cytf and PC.

While all this electron transfer has been going on in Photosystem II, Photosystem I has also been absorbing light, and the pigment P700 has been raised to the excited state where it, too, loses an electron to another neighbouring molecule called X. This molecule, as a result,

Elodea plants live totally submerged in water, and when they are illuminated brightly the oxygen they release in photosynthesis can be seen as bubbles escaping from the leaves.

is raised to a very high energy level. P700 is thus left with a positive charge and must regain an electron before it can again accept energy from the antennae pigments associated with it. The P700 molecule draws this electron from the last member of the chain of substances that passed along the 'spare' electron from molecule Q in Photosystem II – namely PC. This, then, is the point at which the two photosystems are linked. Photosystem I cannot operate unless Photosystem II is providing electrons for it, and Photosystem II cannot operate unless Photosystem I is taking electrons away from it!

When the substance X acquires an electron from the excited Photosystem I pigment P700, it is raised to a very high energy level and proceeds to lose its newly gained electron, and hence its extra energy, through a chain of four different substances in a way similar to that described for Photosystem II. The electron is first passed to a substance FRS, then to FD and then to R. At this point a critical reaction occurs because R passes on the electron to a substance called NADP, which acquires a negative charge and as a result attracts a hydrogen ion or proton (H^+) to join it to form NADPH, the so-called reduced form of the molecule. This is a relatively stable molecule and its importance is that it has a very large 'reducing capacity' – that is, it can pass on its hydrogen atom to other molecules in chemical reactions. When this happens it is oxidized to NADP again and can immediately return to the photosynthetic process and be reduced once more. NADP is therefore essentially a molecule that shuttles hydrogen atoms around the cell so that they can be used to drive chemical reactions that require reducing power, or which need to gain hydrogen atoms. It is the hydrogen atom from NADPH that is eventually used to reduce CO_2 to carbohydrate.

The two photosystems are located in the thylakoid membrane and it appears that the splitting of water takes place on the inside of the membrane and the eventual reduction of NADP to NADPH takes place on the outside. This leads to the production of hydrogen ions (H^+) on the inside of the thylakoid vesicle and their removal from the outside so a gradient in the concentration of H^+ ions is established across the membrane. This gradient is also built up by another process associated with one of the electron transfer steps associated with Photosystem II. In gaining and losing an elec-

tron it appears that the substance PQ collects a hydrogen ion from the outside of the membrane and transfers it to the inside of the thylakoid vesicle. The main point here is that two processes apparently contribute to the establishment of a marked concentration gradient of hydrogen ions between the inside and outside of the thylakoid vesicle, the concentration inside being much higher than that outside.

Before we can appreciate the significance of this concentration gradient of hydrogen ions we must digress for a moment to describe another molecule that has the function of shuttling energy around – a molecule that effectively packages up energy for use later, in some other part of the cell's biochemical machinery. Animals, including man, use the same molecule for the same purpose, but they cannot make it from the energy of the sun; they have to make it from the food they eat, that is from plants. This molecule is called adenosine diphosphate (ADP) and it can be made to combine with another inorganic phosphate group to form adenosine triphosphate (ATP). This reaction requires a great deal of energy, which is then effectively stored in the ATP molecule and can be released under the appropriate conditions when ATP is once more broken down again into ADP and inorganic phosphate. The ADP can be recycled to form ATP again. ADP is therefore an energy shuttle.

Because the concentration of hydrogen ions (H^+) is higher on the inside of the thylakoid membrane than on the outside, the ions diffuse outwards through the coupling factor (CF), and in doing so they bring about the synthesis of ATP from ADP and phosphate. The process effectively takes the captured solar energy held in the (H^+) concentration gradient and transfers it to the ATP where it is stored.

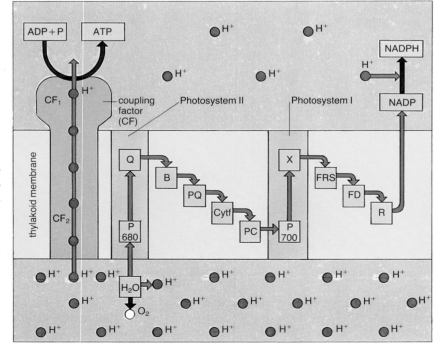

Much of the energy captured in photosynthesis is stored initially in ATP, and the process depends entirely on the concentration gradient of hydrogen ions (H^+) across the thylakoid membrane. Embedded in the membrane is a complex molecular structure called a coupling factor. What is important about this coupling factor (CF) is that because the concentration of H^+ ions on the inside of the membrane is higher than that on the outside, the ions will diffuse outwards through the coupling factor, and in doing so they bring about the synthesis of ATP from ADP and inorganic phosphate.

The result of the two light reactions brought about by Photosystems I and II is therefore to build up in the chloroplast and plant cell a large supply of two essential molecules. The first is NADPH, a molecule with tremendous reducing power, that is the capacity to transfer its hydrogen atom to a wide variety of other molecules while it is itself oxidized to NADP. The second is ATP, a molecule that stores an enormous amount of energy which can be released and utilized to drive other chemical reactions when it is broken down into ADP and inorganic phosphate. Apart from oxygen, these packages of reducing power (NADPH) and energy (ATP) are the sole products of the light reactions in photosynthesis; they move to other parts of the chloroplast and can there – in total darkness – bring about all the other reactions involved.

The diagrams below show how a molecule of glucose (*top left*) and one of fructose (*top right*) join together with the elimination of a molecule of water to produce sucrose, the familiar sugar we buy in the supermarket.

The dark reactions

The two chemical substances ATP and NADPH provide the energy and reducing power required to drive a variety of other chemical reactions, which effectively repackage the energy into a form that can be stored over long periods of time and that can easily be transported within the plant to wherever energy is required for growth and development. In addition, the plant requires carbon, oxygen and hydrogen molecules with which to build up its cellulose supporting structures. The dark reactions satisfy these two principal requirements, first by repackaging energy into a readily transportable and usable form, and second by increasing the amount of carbon, oxygen and hydrogen available to the plant by synthesizing carbohydrates in the form of sugars and starch.

Sugars are complex molecules built up from carbon, oxygen and hydrogen atoms. The number of carbon atoms in different types of sugar molecules can vary from three to seven; the two most common ones, glucose and fructose, each have six and are called hexoses. The glucose molecule consists of a ring of five carbon atoms and one oxygen atom with the sixth carbon atom raised on an arm above the plane of the ring. Fructose, on the other hand is a ring of four carbon atoms and an oxygen atom with carbon atoms numbers 1 and 6 held on arms on either side of the ring. Incidentally, sucrose, the sugar we buy in the shops, consists of a double molecule, a disaccharide, comprising one molecule of glucose and one molecule of fructose joined together by an extra oxygen atom. It is in this form that sugars are transported and often stored in plants.

While the common sugars have six carbon atoms in their molecule, some others have five. These are called the pentoses, and they include xylose which makes an important contribution to the structure of cell walls. Other plant sugars have three, four or seven carbon atoms. These sugars do not accumulate, like glucose and fructose, but they do take part in the dark reactions of photosynthesis and in respiration (Chapter 12). They do not necessarily have a ring structure like glucose and fructose; indeed the three-carbon sugars, the trioses, can not form rings because they do not have enough carbon atoms in their molecules.

In describing the dark reactions of photosynthesis we need only be concerned with the number of carbon atoms in the sugar molecules, and these will be designated C3 for a triose,

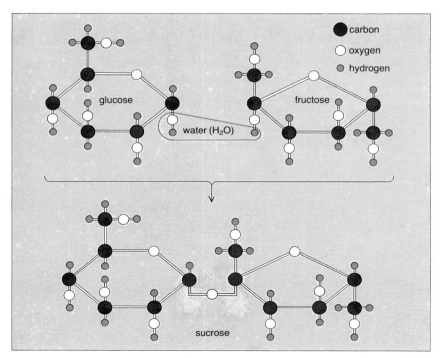

C5 for a pentose and C6 for a hexose. In this way we can concentrate wholly on the way in which carbon dioxide (which has one carbon atom, hence C1) is fixed by the plant and incorporated into sugar molecules. It is necessary to explain the process in this way because plants have evolved two rather different ways of fixing carbon dioxide.

Two major technological developments in the mid-1940s led to our understanding of the dark, carbon-fixing reactions in photosynthesis. The first was the availability for scientific research of radioactive carbon atoms, which could easily be incorporated into carbon dioxide. If a plant was allowed to photosynthesize in the presence of this 'labelled' carbon dioxide, it could later be analysed to establish into which chemical substances the radioactive carbon atoms had been incorporated. By making such analyses at different times after the plant was allowed to begin photosynthesis it was possible to establish the sequence of chemical substances that became radioactively labelled, and hence the sequence of reactions. The second technological advance was the invention of the analytical technique known as paper chromatography. This procedure enables very small quantities of chemical substances to be separated from each other with great purity, and, of course, if some of these are radioactive they can easily be detected and identified.

By using these techniques, and pure cultures of the unicellular alga *Chlorella*, work began which eventually unravelled the complex series of reactions involved in photosynthetic CO_2 fixation. The reactions are, in fact, a cyclical process in which CO_2 is fed in at one point and sugars produced at another.

If *Chlorella* is allowed to photosynthesize in the presence of radioactive CO_2 for a short time, even for only one minute, it is found that many radioactive compounds have already been formed. By reducing the time of photosynthesis to only two *seconds*, however, only one radioactive compound appears in the alga, and this is a three-carbon chemical called phospho-glyceric acid (PGA). So, the first product of photosynthesis is a molecule with three carbon atoms, one of which we know has come from the atmospheric CO_2 because it is radioactive. What then is the nature of the molecule to which this radioactive carbon has become attached? Although it is natural to think it might be a molecule with two carbon atoms, further experiments revealed that it is, in fact,

a sugar with five carbon atoms called ribulose. This sugar has two phosphate groups attached to it and is therefore called ribulose bisphosphate (RUBP). The first reaction of CO_2 fixation in photosynthesis thus involves the combination of a C5 sugar (ribulose) with a C1 molecule (carbon dioxide) to produce an unstable C6 molecule which has never been isolated or identified because it immediately breaks down to produce two C3 molecules of phospho-glyceric acid. The enzyme that catalyzes the reaction is the extremely important protein, ribulose bisphosphate carboxylase.

It is unnecessary here to give details of how all the other steps in the cyclic process were elucidated, but the cycle itself is illustrated in the adjacent panel. It is called the Calvin cycle after its discoverer, and the reason for showing it in even this amount of detail is to enable the reader to see, first, how its operation is wholly dependent upon the provision of energy and reducing power from the light reactions, and second, how CO_2 is fed in at one point in the cycle and sugars are produced at another. Apart from the fascinating reactions involved in the cycle, it should be noted that three ATP molecules (energy) and two NADPH (reducing power) molecules are required to fix each molecule of CO_2. Because the first chemical product of CO_2 fixation in this scheme is a three-carbon molecule, plants that operate the Calvin cycle for fixing atmospheric CO_2 are called C-3 plants. Common examples are broad bean, tomato and wheat.

For many years it was thought that the Calvin cycle was the way in which *all* plants carried out CO_2 fixation, but in the mid-1960s it was discovered that this was not so. Many plants, especially tropical grasses and plants like sugar cane which normally grow in very high light intensities, use a quite different mechanism. In this, the first product is a molecule in which there are four carbon atoms, not three, and the plants that operate this mechanism are therefore known as C-4 plants, or plants that carry out C-4 photosynthesis. These plants do in fact operate the Calvin cycle, but they have a preliminary mechanism which first fixes the atmospheric CO_2 into a four-carbon molecule and then releases it again in a different part of the leaf where it is fed into the Calvin cycle.

Before examining this preliminary CO_2 fixation in C-4 plants we should perhaps first ask *why* many of the plants that grow in very bright tropical sunlight should have evolved a

This summary diagram of the Calvin cycle shows how carbon dioxide is captured in the 'dark' reactions of photosynthesis, how sugar and starch are produced, and how the primary carbon dioxide acceptor RUBP is regenerated. Each purple segment represents one carbon atom, so the number of carbon atoms in the different molecules can be followed easily. The addition of the phosphate groups to the molecules is not shown. Three CO_2 molecules are shown being fixed, with the production of one half-molecule of sugar (3 carbon atoms) and the regeneration of three molecules of RUBP in each turn of the cycle.

The superb detail of an electron microscope photograph reveals the different types of chloroplasts in the mesophyll cells and the bundle sheath cells of the C-4 plant *Zea mays*. The chloroplasts of the mesophyll cell have grana but no starch granules, while those of the bundle sheath cells lack grana but contain conspicuous starch grains.

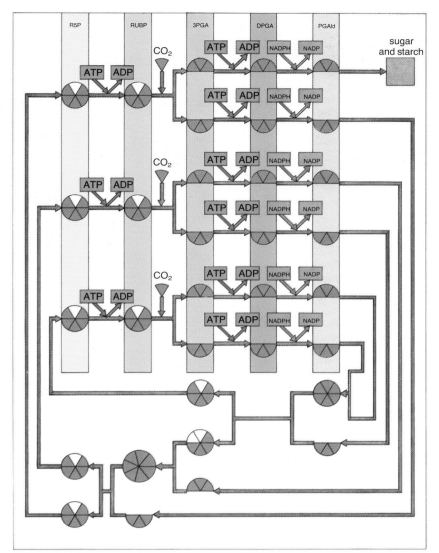

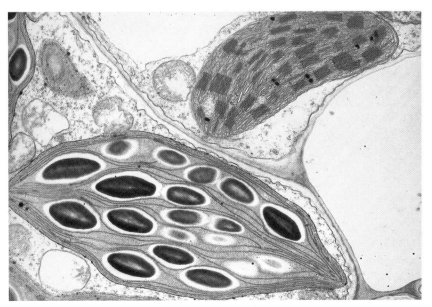

different mechanism of CO_2 fixation from those that grow in temperate regions. The answer seems to be because of a process known as photo-respiration. This process occurs in bright light in plants operating only the Calvin cycle, that is, in C-3 plants, and is due to the fact that ribulose bisphosphate carboxylase, the enzyme which normally catalyzes the reaction between ribulose bisphosphate and CO_2 to provide two molecules of phospho-glyceric acid, can operate in another way in bright light. In these conditions it can add *oxygen* rather than CO_2 to the ribulose bisphosphate molecule, and when it does this the (C5) ribulose bisphosphate is broken down into a two-carbon molecule phospho-glycolate and the three-carbon molecule phospho-glyceric acid. The phospho-glycolate then becomes involved in a series of reactions, the end result of which is that a molecule of CO_2 is *released*. The process thus results in the *loss* of CO_2 from the plant and so the net effectiveness of photosynthesis is seriously reduced. In this way, photo-respiration seriously reduces the productivity and yield of C-3 plants in agriculture and horticulture, and the effect is greater the higher the ambient light intensity. It has been estimated that C-3 plants may lose up to half the CO_2 they fix during photosynthesis by the process of photo-respiration! What useful purpose is served by photo-respiration has not been established by plant scientists, but its control and limitation in many crop plants could produce a very substantial increase in yield, and hence in food supply.

So how is carbon dioxide actually fixed in C-4 photosynthesis? The first important feature of C-4 plants is that their leaf anatomy is different from that of C-3 plants. Around the vascular bundles in their veins they have a layer of special cells which form a 'bundle sheath'. These cells are quite different from the rest of the internal (mesophyll) cells of the leaf in that their chloroplasts have very large starch grains in them and usually lack, or have ill-defined, grana. The mesophyll cells, on the other hand, have normal grana consisting of thylakoid vesicles but they lack starch grains. In some C-4 plants the bundle sheath consists of a single layer of cells, in others it can be two layers of cells thick. This type of leaf anatomy is called Kranz anatomy from the German word for wreath. Most C-4 plants appear to be tropical monocotyledons, such as sweetcorn, sugar cane and sorghum, but many dicotyledonous species also have this type of photosynthesis.

The mesophyll cells and bundle sheath cells carry out different stages in the process of carbon dioxide fixation. The mesophyll cells, with large air spaces between them, can fix atmospheric CO_2 very effectively because they possess large amounts of an enzyme called PEPCase. This catalyzes a reaction that joins CO_2 on to a three-carbon acid to produce a four-carbon organic acid called oxalo-acetic acid, and this acid in turn is quickly converted into two other four-carbon acids called malic acid and aspartic acid, the proportions of which vary from species to species.

For simplicity we shall consider that the only product of this process is malic acid. (This is probably quite true as there is evidence that the aspartic acid is subsequently converted to malic acid anyway.) The CO_2 is thus initially captured and stored as malic acid by the mesophyll cells in a purely dark reaction. The malic acid is then transferred to the bundle sheath cells, which are tightly packed together with scarcely any intercellular spaces. Here, the malic acid is broken down under the action of another enzyme called malic enzyme, and the products of this reaction are CO_2 and pyruvic acid. The latter has three carbon atoms and returns to the mesophyll cells where it is recycled and used to capture another molecule of CO_2. The CO_2 released in the bundle sheath cells is, however, fed straight into the Calvin cycle where it combines with ribulose bisphosphate to form two molecules of the three-carbon phospho-glyceric acid, just as it does in C-3 plants. The bundle sheath cells have a high level of the enzyme RUBP carboxylase which brings about this reaction, and little or no PEPCase, whereas in the mesophyll cells the situation is reversed.

C-4 plants do not appear to carry out photorespiration when exposed to high light intensities and temperatures, and so do not lose any of the CO_2 they have fixed photosynthetically. This makes them very efficient crop plants, and it has been estimated that under similar conditions they can be up to twice as efficient as C-3 plants in fixing CO_2 and producing stored carbohydrates.

Sucrose, starch and cellulose formation

The final energy-storage products in most plants are the disaccharide sucrose and the polysaccharide starch. In addition, much of the

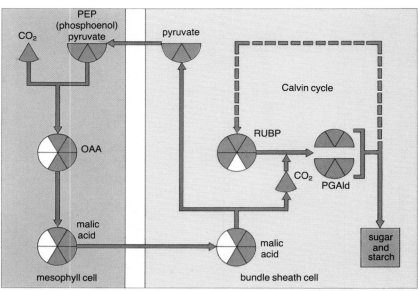

sugar synthesized in photosynthesis is used to form cellulose and other molecules that make up the solid, skeletal structure of the plant – that is, the cell walls. Cellulose, like starch, is made of long chains or polymers of glucose molecules, but in starch and cellulose the glucose molecules are slightly different. This difference is very significant, however, because starch is easily digested by animals and broken down by plants to release its stored energy, whereas cellulose is not. Cellulose is broken down by fungi and bacteria, the latter especially in the stomachs of ruminant animals such as cattle and sheep, with the result that the glucose molecules become available to the animal. It is important to remember that without these bacteria in their stomachs, cattle, sheep and other ruminants would be unable to survive as they would have no means of digesting the cellulose that makes up the bulk of their food.

Sucrose

Sucrose is the carbohydrate energy-storage product that is transported around the plant, in the phloem, from the leaves to the growing points or storage organs. It is formed in the cytoplasm of the cell and so the initial products of the Calvin cycle have first to be exported from the chloroplast.

Some molecules of the phospho-glyceraldehyde produced by the Calvin cycle become changed slightly into molecules that each contain three carbon atoms with one phosphate group attached to them. They are caused to react with one another so that each pair combines to form a six-carbon sugar molecule (a

In C-4 photosynthesis the initial fixation of CO_2 from the atmosphere takes place in the mesophyll cells of the plant's leaf. Here, atmospheric CO_2 is combined with a 3-carbon atom called PEP to make a 4-carbon acid which is immediately converted to malic acid. The malic acid is then passed across to the bundle sheath cells where it is broken down again to pyruvic acid and CO_2. The acid is passed back to the mesophyll cell to be used again, while the CO_2 is fed into the Calvin cycle and converted into sugar and starch.

hexose) with one phosphate group attached to it, while the other phosphate group is released into the cytoplasm. The new six-carbon molecule is called fructose-6-phosphate – the 6 indicating that the phosphate is joined on to carbon atom number 6. Some of these molecules are converted to other six-carbon sugar molecules, particularly glucose-6-phosphate and glucose-1-phosphate which differ from each other only in the carbon atom to which the phosphate group is attached. The molecules of fructose and glucose required to make a sucrose molecule are thus available in the cell; all that is required is the right enzyme to bring about the reaction.

When the energy stored in sucrose is required, the double sugar molecule has to be broken down again into its component molecules glucose and fructose, because only in that form can the sugars enter into the full energy-releasing process known as respiration (Chapter 12). The splitting of the sucrose molecule is achieved by a special enzyme called invertase.

Starch and cellulose

Both of these substances are long chains of glucose molecules, but the way they are joined together differs slightly because the molecules of glucose in the two chains are very slightly different. Molecules of α-glucose, which polymerizes to give starch, are joined between their No. 1 and No. 4 carbon atoms (see illustration). This joining together first requires the combination of the glucose with a phosphate-containing molecule – one of a number of such molecules that provide additional energy for chemical reactions. The 'activated' glucose molecules are then joined on to the starch chain by an enzyme called starch synthetase. The number of glucose units in a starch chain can be between 2,500 and 50,000. The chains are spiral in shape and can be linked by other chains joining on to carbon atom No. 6 of any of the glucose units.

Starch is often formed in chloroplasts in bright light, but it is also formed in stems and roots, sometimes in large, swollen, specialized storage structures like potatoes (stems), carrots and turnips (roots) or bulbs (leaves), where it is formed from the sucrose transported into the organ. The starch in such storage organs occurs in the form of clusters of grains enclosed in membranes. The structures are called amyloplasts, and as we have seen in Chapter 7,

the amyloplasts in root caps and stem bundle sheaths play an important role in the detection of gravity by plant organs.

Cellulose is a polymer of β-glucose which differs from α-glucose only in the arrangement of the H and OH groups on carbon atom No. 1 of the molecule. They are joined together in exactly the same way as in starch except that every other molecule is inverted. This gives rise to the perfectly straight molecular chain characteristic of the cellulose polymer. It is the basic structural material of all plant cell walls, and the chemical molecule underlying the international timber, paper and wood-products industries.

In this chapter we have seen how the energy of the sun's rays is captured and stored in the form of chemical energy in carbohydrates. The process involves the breakdown of water, the release of its oxygen and the use of its hydrogen to drive the reactions that fix and reduce atmospheric carbon dioxide to sugars. These sugars are then transformed and joined together in various ways so that the principal products are sucrose, starch and cellulose. All other chemical substances found in plants are synthesized from the intermediate or final products of photosynthesis in a huge variety of reactions catalyzed by highly specialized enzyme systems.

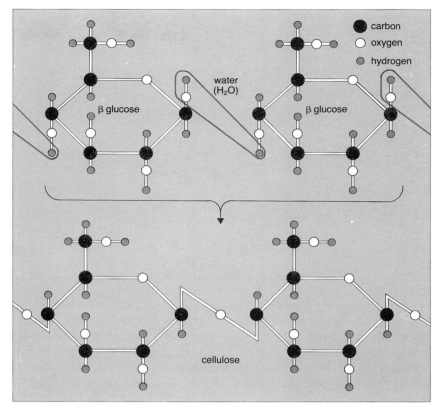

One of the most important products of CO_2 fixation is cellulose – the main component of the plant's cell walls and therefore effectively the 'skeleton' that supports the plant. The diagram above shows how molecules of β-glucose join together by simply eliminating a molecule of water. By this process of 'polymerization' the sugar can build up chains of cellulose many thousands of molecules long.

CHAPTER 10

Plumbing and Pipework

In order to transport water, nutrients, hormones and waste products from one part of the body to another, animals have evolved a complex circulatory system consisting of arteries and veins through which a carrier solution, the blood, is pumped by the heart. Plants have almost identical basic requirements, plus many special requirements of their own, and they too have evolved a highly sophisticated transportation system. In fact, in plants there are two quite distinct systems comprising, for the most part, parallel sets of pipes which run through the plant rather as the hot and cold water pipes run side by side through a house. But there the similarity ends. The hot and cold domestic water pipes are essentially identical, but the plant's two pipework systems work on completely different principles. One system is a positive pressure system, in which the carrier solution is *pushed* through the pipes: the other is a negative pressure system in which *suction* is the driving force. And since the plant has no obvious central pump to do the job of an animal's heart, we are faced with the question of *how* the contents of the two pipework systems are moved along the pipes from one place to another.

The distances involved in some plant plumbing systems are enormous when compared with those of the animal world. Consider a tree in summer-time. The green leaves of the canopy twenty metres or more above the ground are capturing the energy of the sun and converting it into sugar. Water is essential for this process, but the nearest water is in the soil far below. It must, therefore, be absorbed by the roots and then raised twenty metres into the air – by a system with no moving parts! By the same token, for the roots to grow and develop, and absorb the mineral nutrients needed by the tree, the roots themselves need energy. And this energy, in the form of sugar, has to be transported all the way down from the leaves where it is being manufactured.

In order to work out how these transportation systems operate, we first need to see inside the plant and explore the internal structures of roots and stems. To do this, thin slices of the organs are prepared and placed in a chemical solution that stains the cellular structures with characteristic colours. The slice, or thin section as it is called, is then examined under the microscope. Viewed in this way it is immediately clear that the body of the plant is made up of several different types of cells, and that cells of the same type tend to occur in organized clusters forming specialized tissues. By reducing the thin section to a slice barely more than a single cell thick, the structure of the cells can be examined under an electron microscope – at magnifications of 50,000 times life size.

Under an ordinary optical microscope a cross-section of the stem of a herbaceous, broad-leafed plant reveals conspicuous groups of cells arranged in a ring. The groups are called vascular bundles, and they consist of long strands which run the entire length of the stem. The component tissues of the vascular bundles also extend down into the roots where they are rather differently arranged – a subject to which we will return later.

In the stem, the inner part of each vascular

These majestic redwoods (*Sequoia sempervirens*) have their roots in the damp forest soil and their food-producing leaves 80 metres above the ground. In between is a two-way fluid transport system working year-in, year-out, perhaps for a thousand years, without servicing, without any moving parts, and with no source of power but the rays of the sun.

Right: a cross-section of the hollow stem of a buttercup (*Ranunculus* sp.) – a typical dicotyledon. The darker areas embedded in the mass of thin-walled cells are the vascular bundles, shown in the cutaway diagram below.

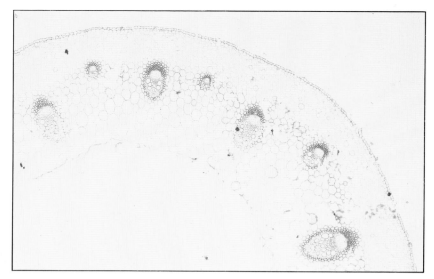

A single vascular bundle (*below*) consists of thick-walled water-transporting xylem cells in the inner half (here stained pink) and thin-walled phloem cells in the outer half (blue), strengthened and supported by a sheath of red-stained fibrous cells.

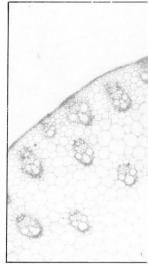

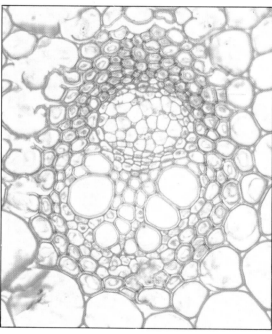

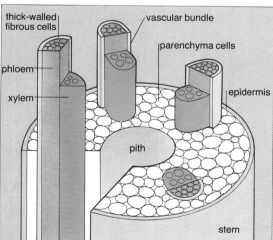

bundle consists of very thick-walled cells forming a specialized tissue called xylem. It is concerned solely with the movement of water and very dilute solutions of inorganic ions upwards from the roots to the shoots. The outer part of each vascular bundle consists of very thin-walled cells, sometimes accompanied by thick-walled fibrous cells which occur at the outermost edge and help to strengthen the stem. The thin-walled cells make up a tissue called phloem, and it is this tissue that transports a solution of organic compounds such as sugars, amino acids and hormones both upward and downward through the stem.

The vascular bundles in a herbaceous plant stem are for the most part embedded in a mass of more or less spherical cells which expand under the effect of their own hydraulic pressure-generating system (Chapter 6). When fully pressurized, these cells press tightly against each other and against the epidermis, the outermost skin of the plant stem, and so provide most of the stem's rigidity and mechanical strength. If these simple, unspecialized body cells, or parenchyma, lose their hydraulic pressure through shortage of water the plant loses its structural strength and starts to wilt.

The same two specialized types of transport tissue are found in the roots of herbaceous plants, but here their physical arrangement is rather different. The phloem strands in the root are located between the projecting 'spokes' of a xylem core, which may or may not have a central zone of parenchyma cells. The central complex of vascular tissue is surrounded by two most important layers of cells, an inner

The sweetcorn (*Zea mays*) stem illustrated above is typical of monocotyledons: the vascular bundles are numerous and are scattered throughout the stem with no regular arrangement.

Higher magnification of the sweetcorn stem (*right*) shows the 'monkey face' appearance of the vascular bundles which is typical of grasses and cereals.

Very high magnification (*right*) shows the vascular bundle to consist of a sheath of small, thick-walled cells (yellow), three large xylem cells and one smaller one – the lowermost of these cells having been damaged by growth of the surrounding cells – and thin-walled (purple) phloem tissue made up of large octagonal sieve cells and smaller, darker, companion cells.

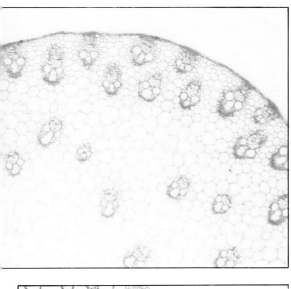

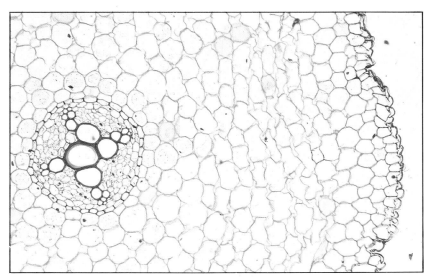

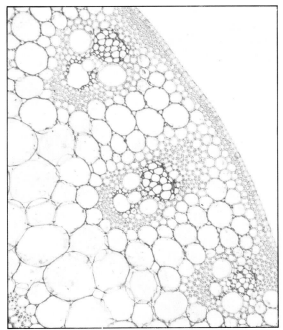

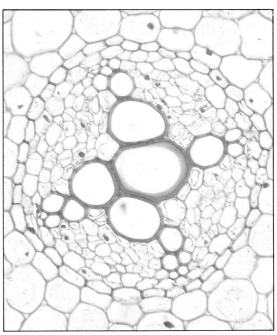

The stained cross-section of a buttercup root above shows the arrangement of cells typical of dicotyledons. Large thin-walled parenchyma cells (blue) make up the cortex – the main body of the root – at the centre of which is a core of much smaller cells called the stele.

Higher magnification (*left*) reveals that the stele is surrounded by a layer of thin-walled cells (the pericycle) and then by a layer of thick-walled cells (the endodermis). The stele itself consists of a four-spoked core of xylem cells (stained red) with four separate groups of (blue) phloem cells lying between the spokes.

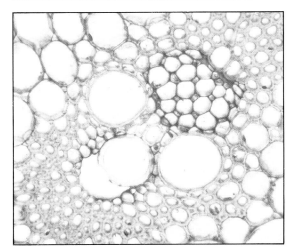

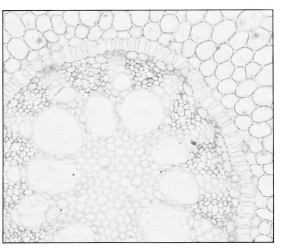

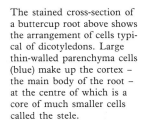

Monocotyledon roots such as those of the iris (*left*) are similar to dicot roots except that there are many more 'spokes' of xylem, often around a core of thin-walled parenchyma cells.

pericycle and an outer endodermis, and then by a rather less interesting multilayered cortex, or covering, of parenchyma cells. The importance of these tissues will be described in due course.

Let us now take a closer look at the cellular structure of the xylem and phloem, because if they do comprise a system of pipes then we can at once start to consider how the plant's internal transport systems operate.

Xylem: the plant's water transport system

The xylem consists of two types of cells called tracheids and vessels, both of which form pipes through which liquid can be moved. One of the most interesting features of these structures is that once the individual cells have reached their predetermined size and form they promptly die. The cell content, the protoplasm, then disappears to leave the thick cellulose cell wall forming a rigid empty pipe through which water can move without interruption. The living plant's xylem pipework thus consists entirely of dead cells.

Most of the tracheids in a plant stem are known as 'pitted tracheids'. They are elongated cells with thick strong walls, and where they are joined to their neighbours there are small holes, or pits, in the walls so that the cell cavity is connected with the interior cavities of neighbouring cells above, below and at either side. A strand of tracheids thus forms a series of pipes along the stem, with constrictions at the holes in the walls where two cells make contact. These constrictions must increase the resistance of the pipe to the flow of water. In vessels, on the other hand, the pipes are much wider because the cells themselves are much broader, and there is much less obstruction of the pipe where individual cells meet their neighbours above and below. This is because the vessel cells develop in vertical columns along the shoot or root, and where two cells meet their transverse end walls are completely removed during the final stage of development, before the cells die. A vessel thus comprises a relatively large-diameter pipe, free of any constrictions. As will be explained later, wood consists entirely of tracheids and vessels, which gives some idea of the tremendous strength of these cells. One of the most important building materials in the world is merely the redundant plumbing of a tree!

We must now consider what makes the water move through this system of pipes. The tremendous thickness of the cell walls has already provided an important clue. As water evaporates from the leaves it is replaced by water drawn up through the xylem tubes. This means that the continuous columns of water in the tubes literally hang from the top. They are pulled upwards, against the force of gravity,

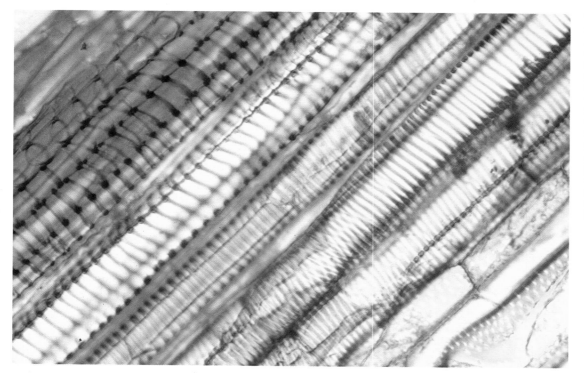

This section along the stem of a sunflower shows the various types of tracheids that occur in different parts of the xylem. At the far left is the first-formed xylem, in which the tracheids have rings of wall thickening. To their right lie several columns of tracheids with spiral wall thickening. Both types can stretch to accommodate growth. At the right of the photograph are pitted tracheids which can not stretch: they were formed after this layer of stem cells had finished growing.

This apparatus, called a potometer, measures the amount of water taken up by a plant – in this case a cut *Pelargonium* shoot. The rate at which the air bubble moves along the graduated glass tube gives a precise measure of the volume of water (coloured red for clarity) absorbed per minute, and this is roughly equivalent to the rate of water loss (transpiration) by the leaves.

by suction at the upper end. This immediately explains why the cell walls are so thick: they have to withstand the enormous negative pressure created by the suction and by the water columns suspended inside them under tension. Everyone is familiar with the problems of sucking lemonade up a drinking straw. Suck too hard and the straw collapses, preventing any movement of the liquid. Were it not for their great strength, the xylem tubes would simply collapse under the inward pressure: no water would flow up the stem and the plant would be unable to draw in water through its roots.

But how much tension can a column of water withstand? The tension in a 100-metre-high redwood tree, for example, must be enormous – especially in hot dry weather when soil water is in short supply. The tensile strength of a water column is actually more than adequate to support a column to the top of the tallest trees known, indeed it must be so, otherwise such huge plants could not exist. But the columns do break in dry weather when the tension becomes abnormally high, and they make a distinct cracking sound as they part. Just how the water columns repair themselves is not known, but it is very unlikely that more than a very small proportion of the total are broken at any one time.

Since the xylem pipes are not open at either end, we must next consider how water is lost from the leaves and how replacement supplies are taken in through the roots.

The underside of the leaf, and in some plants the upper surface too, is dotted with tiny pores called stomata (Chapter 9) which have the ability to open and close according to the

prevailing conditions. With the pore open, the atmosphere inside the leaf (that is, in the network of spaces between the individual leaf cells) is in direct contact with the external atmosphere. If the outside atmosphere is less than 100 per cent humid, water will be lost by evaporation from the interior of the leaf, through the stomata. Even an external humidity of 99 per cent has considerable potential to draw water out of a leaf. The water thus lost from *between* the leaf cells is replaced by water evaporating from the minute spaces *within* the cellulose walls of the leaf cells, and this in turn is replenished by water drawn through the cell walls of the xylem tubes which extend through the veins of the leaf. The net result is that water is drawn up the xylem tubes by evaporation losses from the leaves. At the beginning of spring there is, for a short time, a positive pressure in the xylem. It is called 'root pressure' and is necessary to 'prime' the system by getting water up to the buds and new leaves. However, as soon as the buds open, the leaves start to create the normal negative pressure conditions and the root pressure rapidly disappears.

Considerable force is required to move the water column upwards since both the weight of the column and the resistance to flow in the narrow tubes must be overcome – as well as the resistance to entry at the bottom end of the system. Here, water has to be drawn into the xylem tissue from the soil through the root cortex. And it is at this point that the importance of the root endodermis comes to light. It is possible for water to move by simple capillary action through the cell walls and intercellular spaces of the cortex and so make its way from the soil towards the interior of the root without actually entering the protoplasm of the cortical cells at all. But this purely physical movement is brought to an abrupt halt at the endodermis because the radial cell walls of this layer are completely impermeable to water. The only way that water can continue through the endodermis and into the xylem beyond is by passing through the living protoplasm of the endodermal cells, and in particular through some rather special endodermal cells called 'passage cells'.

As we have seen in Chapter 6, the cell protoplasm is surrounded by a highly specialized semipermeable membrane which can keep out particular ions and chemical compounds. But if, on the inner side of the endodermis, there is a strong negative hydraulic pressure –

a suction – emanating from the xylem tubes, water will be drawn through this living filter mechanism. The endodermal cells are, in fact, the only living cells through which water *must* pass on its long journey from the soil to the leaves far above. For the rest of its journey it passes through the minute spaces between living cells and through the main pipework system of the xylem tissue.

Plumbing on the grand scale

So far we have been concerned primarily with herbaceous plants. But in much larger plants such as trees, where the leaf area is huge and the water losses enormous, there is need for plumbing on a much larger scale in order to cope with the huge quantity of water that must be carried up the trunk every day. A mature silver maple, for example, may lose up to 265

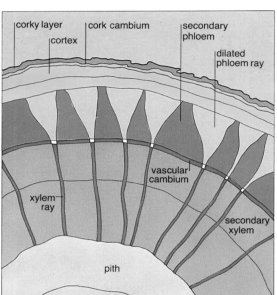

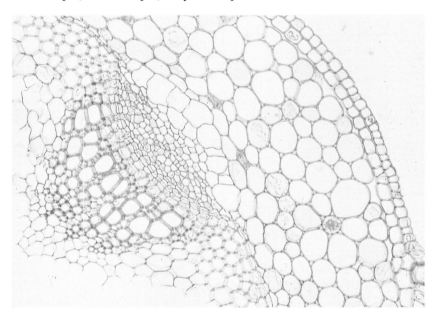

Secondary thickening of dicotyledon stems starts with the onset of cell division in the vascular cambium (*above* and *top left*) – a ring of cells that passes between the xylem and phloem of the original vascular bundles.

A three-year-old lime stem (*far left*) has a small core of thin-walled pith cells surrounded by a thick layer of secondary xylem arranged radially and showing three clear annual growth rings. Outside this lies the secondary phloem, in which are embedded a large number of thick-walled fibrous cells (here stained purple). The medullary rays (*left*) pass through the vascular cambium and spread out in the phloem layer beyond.

litres of water an hour in the later afternoon of a warm summer day, and every drop must be replaced from the soil below if the leaves are not to wilt.

In trees, of course, there is an additional problem; that of providing enough mechanical strength to support such a massive structure. Both the additional pipework and the structural strength are provided in the form of a massive amount of new xylem, produced by a process known as secondary thickening. In a tree the xylem pipes are essentially the same shape and size as those in the stem of a nettle, a buttercup or any other herbaceous plant; there is just an enormous increase in the number of pipes, and so in the plant's capacity to transport water.

Secondary thickening is a simple process to understand. The primary or herbaceous type of structure of the very young broad-leafed or dicotyledonous tree seedling is later modified by the development of a cylindrical layer of actively dividing cells called a vascular cambium. This cambial ring begins its development between the xylem and phloem cells in the primary vascular bundles. It then spreads out on either side until it links up with the cell layer reaching out from neighbouring vascular bundles. Once the ring is complete, the cells of the vascular cambium continue to divide, producing layer upon layer of new cells on both the inside and the outside of the ring. The cells on the inside develop into vessels and tracheids while those on the outside develop into phloem. The vascular cambium lasts for years – for hundreds of years in the case of oak and redwood trees – and each year it produces more new xylem cells on its inside surface and more new phloem cells on its outside. The hard trunk of a tree thus grows in diameter, the new xylem of each year's growth providing additional capacity for the plumbing system. The old xylem, deeper inside the tree, gradually becomes blocked up and undergoes various chemical changes. From that point on it no longer functions as part of the plumbing system but instead provides the support necessary to keep the tree upright.

The annual growth rings familiar in the cross-section of a felled tree reflect the relative number of vessels and tracheids formed from the cambium cells at different times of the year. Vessels, being much larger in diameter than tracheids, allow a much greater flow of liquid up a plant stem. They tend to be formed in early spring and summer when bud break and the expansion of new foliage creates a huge demand for water in deciduous trees. Later in the season, as the leaves become old and non-

Wood is the secondary xylem of a tree, and the different types of cell give different woods their characteristic properties. The micro-photograph *below left* shows large vessels and small tracheids in a section of the trunk of *Robinia pseudoacacia*.

A vertical section of the same wood, looking from the outside towards the middle of the tree (*left*) reveals a pitted vessel made up of several elements joined end to end. The small spherical cells belong to one of the medullary rays, and the long thick-walled cells are tracheids.

The detail below shows an old vessel in *Robinia* wood to illustrate its construction from a number of barrel-shaped vessel cells joined end to end. The thin, inwardly-bulging structures are tyloses, and they eventually block up the vessel. In some trees these contain chemicals that inhibit fungal growth – producing very high quality rot-resistant hardwood timber.

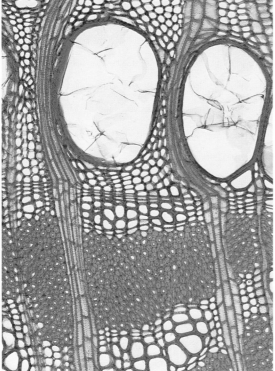
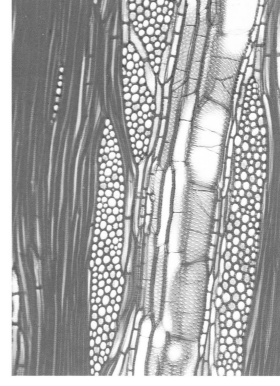
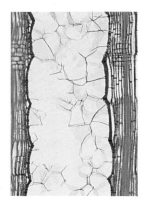

functional, the demand for water is greatly reduced and may indeed be restricted by the formation of tracheids only in the stem. Vessels occur only in angiosperms, the flowering plants: they do not occur at all in gymnosperms such as firs and pines. Annual rings do form in these coniferous trees, but they are due to the different sizes of tracheids formed in the spring and autumn. The spring rise in water movement in evergreen trees is much less dramatic than in deciduous trees – largely because the evergreens retain a large number of functional leaves throughout the year, adding to them each spring with a flush of new growth.

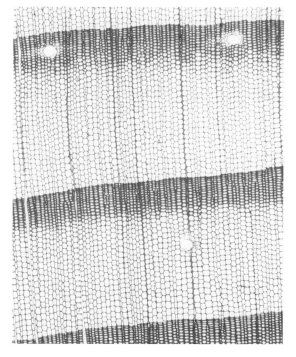

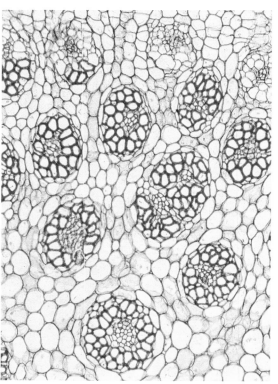

The stem of a pine tree (*top left*) is made up entirely of tracheids, which vary in size according to the season and so produce growth rings.

A vertical section of the same wood (*above*) reveals a mass of rather uniform tracheids. Special pits in the tracheid walls show up as small swellings.

In some monocotyledons, such as palms, *Yucca* and (*far left*) *Dracena*, secondary thickening is caused by cells formed on the inner surface of a cambium later differentiating to form vascular bundles.

In the vascular bundles of secondarily-thickened *Dracena* stems (*left*) the tissues are arranged in a most unusual way – with xylem surrounding phloem.

The stem of the doum palm (*above*) has undergone secondary thickening by the processes described in the photographs at the foot of the facing page.

the tremendous hydraulic pressure in the protoplasm, but in the presence of certain hormones it can reduce its tensile strength in a controlled way so that the cell can expand or grow. Only after the cell has attained its final size does it lay down secondary wall material – thickening and strengthening its walls but thereby ruling out any possibility of further growth.

The problem for the tracheids at the very tip of the shoot is that they have to be able to stretch along with their living, growing neighbours while they themselves must be fully developed, dead, functional tubes capable of withstanding a considerable negative (inward) pressure. To meet this set of requirements the cells develop very rapidly, but instead of laying down secondary cell wall material uniformly, all over the inner surface of the primary cell wall, they lay down the secondary material in very specific patterns. These patterns are of two main types. The first and earliest tracheids lay down annular rings of secondary material while the later tracheids lay it down in a spiral pattern, rather like a helical spring. Each cell is, in effect, like a piece of thin plastic tubing, with either a series of hoops or a coiled wire stuck to the inside to provide support and keep the tube open. Such a tube is flexible and can be stretched, but the strong internal hoops or coils prevent its walls collapsing when a negative pressure is applied inside the tube. This beautiful adaptation of tracheid development occurs only in tracheid cells that develop above or in the growing zone of a stem. Those that develop lower down never face the stretching problem and so develop walls that are more or less uniformly thickened, except for the small holes or 'pits' that allow water to pass from one tracheid to another.

Phloem: the nutrient transport system

The phloem also consists of two distinct types of cell, but although these, too, are elongated, they are structurally completely different from those found in the xylem. First, they both have extremely thin walls, whilst the xylem cells are characteristically thick-walled; second, they are living cells – that is, cells that have retained their cytoplasmic contents. The two types of phloem cells are the sieve cells – the principal cells through which nutrients are transported – and companion cells, which seem to be very closely associated with the sieve cells and may indeed be essential to their functioning as a transportation system.

Negotiating the growing zone

In a growing plant, water has to be transported right to the very tip of the shoot otherwise it will die. But this means somehow transporting the water through the zone of rapid cell growth. How can a strong, rigid, pipe survive, undamaged and functional, when the cells all around it are in the process of doubling or tripling their length?

The answer to this apparent impossibility is that the very first xylem tracheids to differentiate, right at the top of the stem, above the growing zone, are of a very special type. They have an amazingly complex cell wall development which allows them, despite being dead, to stretch as the living cells around them grow, thus ensuring the continued supply of water to the shoot apex.

In the zone of cell division, right at the tip of the stem, new cells form only what is known as a primary cell wall. Chemically this consists of carbohydrate polymers, the main one being the polymer of glucose known as cellulose. This primary wall has the ability to withstand

Microscopic examination of the cells in the vascular bundle of a herbaceous plant shows that there is at least one companion cell associated with each sieve cell, and that the sieve cells are joined together end to end to form a continuous tube – the end walls having become perforated at the final stage of development. These modified end walls are quite complex and are known as seive plates. They are formed by adjacent cell end walls developing corresponding and perfectly matched perforations, and the connecting holes thus formed are lined around their edges with a special substance called callose. The living cytoplasm forms a thin layer lining the inside of the cell and is separated from the cell wall itself by a semipermeable membrane, the plasmalemma. The cytoplasm extends from one cell to another through the sieve plate pores. Close examination of the sieve cells, however, reveals that they lack a nucleus – a very strange state of affairs indeed since the nucleus is the repository of most, if not all, of the information required to keep the cell functioning. It may be that the sieve cell has lost its nucleus because such a bulky object in each cell would impede the flow of the nutrient solution, and such a suggestion would be perfectly reasonable providing some suitable arrangement could be made for nuclear functions to be available close by. This, almost cerainly, is where the companion cell comes in: its role seems to be to make available to the sieve cell those metabolic functions that it may have sacrificed in order to function better as a fluid-carrying pipe. The companion cells contain very dense cytoplasm and prominent nuclei and they are, in fact, sister cells to the sieve cells with which they are associated.

In trees and large, woody, perennial plants the secondary thickening process which produces new xylem each year also produces new phloem. Layers of new cells produced on the outer surface of the vascular cambium grow and differentiate into a ring of secondary phloem, again consisting of sieve cells and their attendant companion cells, together with some fibrous support cells and parenchyma. This outer ring of secondary phloem is clearly renewed each year as new phloem tissue is added to its inner surface. The older cells cease to function, and die. The bark of the tree consists of the new and old secondary phloem plus a thick layer of hard and totally impervious cells – cork – which form in the outermost layers of the living phloem from another special

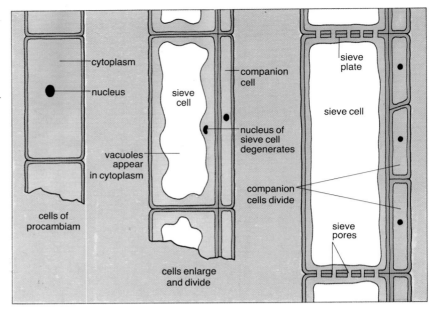

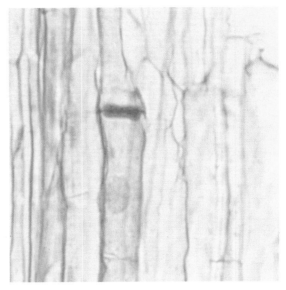

Three stages in the formation of a sieve cell and its companion cells are shown in the diagram above. In the longitudinal section through the phloem of a cucumber stem (*left*) the elongated sieve cells are clearly visible. The darkly stained areas at the ends of the cells are the sieve plates.

ring of actively dividing cells called the cork cambium. This cork layer represents one of the tree's most important defences. As the trunk of the tree grows and expands, the outer layers of old phloem naturally split and crack, and the cells exposed in these fissures – rich in sugars and other nutrients – would be ideal breeding grounds for bacterial and fungal infections. The formation of an impervious cork layer in the ageing secondary phloem layers provides a self-sealing protective barrier against such infections.

We have established that the phloem transport system consists of the elongated thin-walled sieve tubes, but what mechanism generates the flow of fluid through these pipes? One thing is certain: it cannot be a mechanism based on suction, as in the xylem, because the application of a strong negative pressure to one end of a tube with such thin walls would cause its immediate collapse. The structure of the sieve tube is such that it can only function under positive pressure – that is, by a high pressure at one end pushing the fluid through the tube. And this type of system has one very important property, without which many plants would be unable to survive from one year to the next. That property is that the system is easily reversible. Reversal of flow in xylem is theoretically possible, but in practice it is quite impossible because the roots will always be in or near water and the leaves nearly always in an atmosphere less than 100 per cent humid. Water either moves up the stem, or does not move at all. In the case of the phloem system, however, reversal of flow *is* possible because molecules can be loaded into the system at either end – thus creating the increased positive osmotic pressures necessary to generate flow. In the summer months phloem transport in virtually all plants will be downward – carrying nutrients from the leaves to other parts of the plant. In winter, the tree becomes dormant. Its systems are all set to 'tick-over', hardly any fluid movement takes place at all, and the huge plant simply waits for spring – surviving on the nutrients distributed throughout its tissues. But for many perennial plants, survival through the winter and the ability to send up new shoots in the spring depends entirely on energy reserves stored in underground organs such as roots and tubers, rhizomes and bulbs. And for these organs to be able to send their energy supplies to upward-growing shoots, a reversible positive-pressure system is essential.

The fact that the phloem fluid transport system works on positive pressure was initially deduced purely from the structure of the cells, but this mechanism can also be demonstrated very neatly in a number of simple experiments. The ideal subject for these demonstrations is a castor oil plant about one metre high.

If a small cut is made in the outermost part of the stem and then gradually increased in depth, a drop of liquid will suddenly appear on the surface and grow in volume. Clearly the cut has reached tissues containing liquid under positive pressure, and microscopic examination quickly confirms that this tissue is phloem. If the cut is then made deeper still, a point is reached at which the exuded drop of

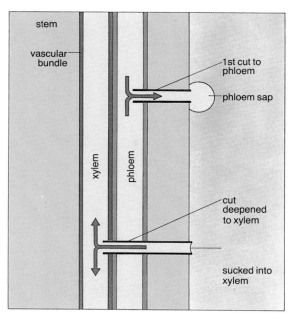

When a cut is made into the stem of a *Ricinus* plant, the drop of phloem sap that oozes out proves that the phloem is under positive pressure. If the cut is made a little deeper, into the xylem, the negative pressure in that system immediately sucks the sap back inside!

phloem sap suddenly disappears. The cut has reached the xylem, and the strong negative pressure in the xylem pipes has sucked the droplet back into the stem!

However, when a cut is made into the phloem tissue, the drop of sap that exudes does not continue to grow indefinitely. The plant appears to have an automatic mechanism for preventing uncontrolled haemorrhaging of its vital nutrient supply. Microscopic examination of the sieve tube shows that the sudden release of pressure caused by the initial cut, and the sudden rapid flow of fluid in the tube, together cause a sealing reaction to take place at the pores in the sieve plate. The damaged cells are therefore very quickly sealed off and the plant's fluid loss is minimized.

Several years ago, quite large quantities of phloem sap were required for chemical analysis at a major research centre, but efforts to obtain them were frustrated by the damage-limitation system described above. For reasons that have never been revealed, the scientist in charge of the project decided one day to rub the plant stem gently up and down over a distance of about ten centimetres above the cut. Amazingly, this gentle stimulation apparently prevented the sealing action from taking place, and the phloem sap continued to flow from the cut at a steady rate. Over a period of several hours, five or more millilitres of the precious fluid could be collected. Thus, a great deal of important analytical work on phloem sap owes its existences to the establishment of the world's first massage parlour for plants!

In extremely humid conditions, phloem sap may even be exuded through leaves – as illustrated here by the leaves of the lady's mantle, *Alchemilla*.

An earlier method of obtaining phloem sap was to allow an aphid to insert its stylet into the plant stem – its normal method of feeding – but then to anaesthetize the insect and quickly cut off the stylet near the insect's head. This left a microscopic tube in place in a phloem cell through which the sap would ooze out very slowly. It was a delicate operation, and yields were very small, but at that time there were few alternative methods.

Analysis of the phloem sap shows it to be a quite concentrated solution of organic compounds with the sugar sucrose, the main product of photosynthesis, present in the greatest amounts. The sap also includes amino acids, the building units for proteins, and other organic compounds including the growth-regulating chemicals. The fact that sugars and amino acids are transported throughout the plant by the phloem system is confirmed experimentally by 'labelling' various compounds with radioactive markers and then monitoring their movement through the plant tissues.

Some readers may wonder whether it is not too dramatic to talk of fluid transportation systems when it might be perfectly possible for sucrose synthesized in the leaves simply to diffuse along the leaf stalk, stem and root along a concentration gradient. This possibility, however, is ruled out by the fact that the speed of transportation along a stem is hundreds of times greater than could possible occur on the basis of simple physical diffusion. Furthermore,

it is clear that living cells are involved in phloem transportation because if the cells in a section of stem are killed, by treating the stem with steam, no phloem movement can take place across the treated zone. Upward movement of water through the xylem, however, remains completely unaffected.

The phloem system therefore consists of thin-walled pipes containing a moderately concentrated solution of sugars under positive pressure. How, then, does the system generate the force needed to bring about fluid flow? A number of mechanisms have been proposed over the years but the one that commands the greatest support from experimental work is the 'mass flow' hypothesis suggested by Ernst Münch in 1927. Mass flow simply means that the fluid or solution moves through the conducting pipes with the solvent and the dissolved chemicals moving along together, at the same speed. To achieve this flow, there must be a greater pressure at one end of the pipe than at the other, and this pressure increase is created osmotically when the plant 'loads' sugar molecules into the phloem cells – either from the chemical factory of its leaves or from storage organs in the ground.

The principle of the mass flow hypothesis is best explained by the physical model illustrated here. In the living plant essentially the same mechanism operates, with the additional complication that to maintain a continuous flow, sucrose must be continuously loaded into the phloem sap at one end (the source) in order to maintain a high concentration, and off-loaded at the destination (the sink) to reduce the concentration there.

Consider the case of a mature tree in summer. Its leaves are actively photosynthesizing and sucrose is being transported down the trunk to the growing roots. At the top end of the system photosynthesis produces large amounts of sucrose which are then, by some process not fully understood, loaded into the sieve cells. Whether or not the companion cells play any part in this process is not known, but it is certain that energy is used in the loading process because work must be done in order to pump the sucrose into the sieve cells against a concentration gradient so as to produce and maintain a high concentration at the upper end of the sieve tube. This high concentration will naturally draw water into the cells by osmosis and so create a high hydraulic pressure in the tube.

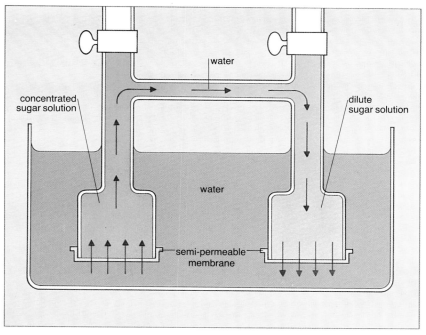

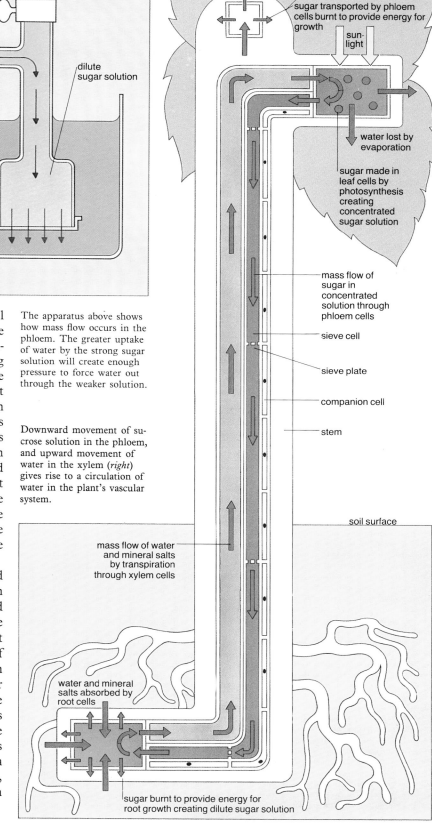

At the bottom end of the system sucrose will be consumed by the growing roots. Since these subterranean organs cannot carry out photosynthesis, they have no means of manufacturing their own energy supplies and must therefore draw all their requirements from the nutrient solution reaching them through the phloem tubes. This active use of the sucrose lowers the osmotic potential of the sieve cell contents and the result is that the hydraulic pressure in the sieve tube forces water out of the tube and into the intercellular spaces. Thus, the input of sucrose, and hence water, at the top of the system and the removal of sucrose, and hence water, at the bottom end creates a pressure gradient and a movement of fluid down the pipes.

Curiously, at least some of the water forced out at the bottom will undoubtedly be taken up by the negative pressure in the xylem and will thus find itself transported back up the tree and into the leaves where it may be lost to the atmosphere, incorporated into new leaf cells, used in photosynthesis or even drawn back again into the sieve cells to begin another downward journey in the phloem. So, despite having two very distinct transportation systems in the xylem and phloem, plants may also be said to have a circulatory system in that it is quite possible for water to be transported in a closed loop between the leaves and the roots, travelling upwards in the xylem and down again in the phloem.

The apparatus above shows how mass flow occurs in the phloem. The greater uptake of water by the strong sugar solution will create enough pressure to force water out through the weaker solution.

Downward movement of sucrose solution in the phloem, and upward movement of water in the xylem (*right*) gives rise to a circulation of water in the plant's vascular system.

115

CHAPTER 11

Mineral Nutrition

Animals require a continuous supply of mineral nutrients if their growth and development are to take place normally. Without them they soon become unhealthy, and may even die. This, of course, is why children are encouraged to drink their milk and eat their green vegetables: the two foods provide a convenient source of their daily requirements of the essential minerals calcium and iron. But plants, too, have specific mineral needs. They soon become sickly and die if they are unable to obtain certain essential minerals, and such nutrient deficiencies severely reduce the yield of agricultural and horticultural crops in many parts of the world.

Minerals are inorganic chemical compounds which occur in various combinations in rocks, and as the rocks are gradually broken down by the natural process of weathering, some of the mineral elements are released. They then react with atmospheric oxygen, carbon dioxide, sulphur dioxide and other gases to produce a wide variety of mineral salts which dissolve in the soil water and thus become available for absorption by the roots of plants.

Inorganic (mineral) salts are essentially those that do not consist primarily of carbon, hydrogen and oxygen. Common salt, sodium chloride, is an inorganic salt; so are copper sulphate, potassium nitrate and iron chloride. When dissolved in water, inorganic salts tend to disintegrate into two components, or ions, in a process called ionization. Each individual ion carries an electrical charge, either positive (cations) or negative (anions), the importance of which will be evident later. For example, potassium nitrate has the formula KNO_3, and in solution it ionizes to produce one potassium ion with a positive charge (K^+) and one nitrate ion with a negative charge (NO_3^-). Similarly, sodium chloride produces sodium cations (Na^+) and chloride anions (Cl^-).

Plants absorb virtually all their mineral nutrient requirements from the soil solution – that is, the film of water around the soil particles – and more specifically they absorb their requirements as ions. Plant roots must, therefore, have some way of extracting these electrically charged particles from the soil solution in which they are present at extremely low concentrations. Furthermore, out of the large number of inorganic ions present in the soil solution, the plant requires at most only about 13, and it requires them at rather higher concentrations than those at which they occur in the soil solution. This means that the roots have to operate a very sophisticated collecting system which actually pumps the required ions into the root cells against a concentration gradient. This clearly requires work to be done – so energy must be supplied to the pumps. In addition, there must be a system for identifying the ions required and for rejecting those that are not. The ion pumps are therefore not just general ion pumps but *selective* ion pumps, which bring about the accumulation in the plant of the 13 or so ions that are essential for its healthy growth and efficient operation. The fact that plants absorb and accumulate selected ions in this way explains why they are such a rich and valuable source of mineral nutrients for animals.

The elaborate pitchers of the *Nepenthes* pitcher plant are formed on highly modified projections of the plant's foliage leaves (*inset*). Inquisitive insects attracted to the trap attempt to alight, but lose their grip on the waxy rim and plunge to their death in the enzyme solution below. There, the body tissues are digested to provide the plant with nitrogen.

One inorganic nutrient, nitrogen, can be absorbed from the soil in the form of nitrate ions (NO_3^-) or as ammonium ions (NH_4^+), but in many plants it is acquired directly from the atmosphere, which is about 80 per cent gaseous nitrogen. Nitrogen is one of the most inert or non-reactive gases known, but quite a large number of plants have evolved a way of capturing the gas from the atmosphere and converting it to ammonium ions. This process almost always involves setting up a special relationship with another organism, usually a bacterium or a related organism called *Frankia*, although some of the primitive algae and bacteria can carry out this process alone. The fixation of atmospheric nitrogen will be described later in some detail because it is of such immense importance to agriculture, especially in developing countries where the cost of nitrogenous fertilizers is far beyond the means of most farmers.

The thirteen essential mineral nutrients

There are two principal ways of determining which inorganic nutrients are required by plants. The first is to make a detailed analysis of the inorganic substances present in a plant. This will reveal which ions the plant absorbs from the soil, but it will not necessarily indicate which of these are *essential* for the plant's growth and development. The second method is to grow plants experimentally with their roots in solutions containing mixtures of highly purified inorganic chemicals from which particular ions have been excluded. If the plant grows normally, then the ion excluded is not an essential nutrient; if the plant looks sick, it clearly is. Chemicals of great purity are required for such studies, and great care has to be taken that only water that has been distilled many times, and containers that have been thoroughly and repeatedly soaked and washed in such water, are used. These extreme precautions are necessary because some inorganic ions are required in such minute amounts that they can be adequately supplied from traces on the walls of glass containers even after they have been washed several times!

Such studies have shown that the main inorganic elements essential for healthy plant growth are nitrogen, phosphorus, potassium, calcium, magnesium and sulphur. These elements are required in appreciable quantities and are therefore known as the macronutrients. Several other inorganic elements are also essen-

tial for healthy growth and development, but these are required in extremely small quantities and are therefore called micronutrients. This group consists of iron, chloride, copper, manganese, zinc, molybdenum and boron.

In addition to these 13 essential nutrients there is, of course, a requirement for the three basic building blocks, carbon, hydrogen and oxygen, which are acquired by the plant from atmospheric carbon dioxide, oxygen and water. All plants require these 16 inorganic elements, but some plants also have a special requirement for a further element. Sodium, for example, is essential for the health of the halophytes, the plants that grow in salt marshes.

The special roles of the nutrient minerals

We have already seen, in Chapter 9, that carbon, oxygen and hydrogen are required by plants to make the carbohydrates produced in photosynthesis. The photosynthetic end-products, and many of the intermediary compounds in both the C-3 and C-4 types of photosynthesis, enter into a large number of chemical reactions. In some of these nitrogen is added to the molecules to make amino acids, which are the building units of proteins and enzymes. Nitrogen is also incorporated into nucleic acids, chlorophyll, many co-enzymes and some plant growth hormones. Some amino acids contain a sulphur atom, and these are essential for the manufacture of certain enzymes. Phosphorus is essential for the manufacture of the energy carrier ATP, and is also an essential component of nucleic acids, while phosphate groups must be added to many sugar molecules before they can enter into some of the reactions essential

Sea purslane is a common halophyte found in the salty environment of sea shores and estuaries in Europe, the Mediterranean and north and south Africa.

118

for photosynthesis and respiration. They are also involved with fat molecules in forming the phospholipids that are vital components of cell membranes.

Calcium is an important component of plant cell walls and particularly of the pectin material that actually sticks the cells together. Calcium pectate is one of the most insoluble substances in nature, and is thought to play a critical role in this adhesive process. Calcium is also important for the maintenance of the semipermeable properties of cell membranes. In the absence of calcium, the membranes become permeable to a range of molecules with the result that the cells' hydraulic pressure-generating mechanism can not operate. Magnesium is an essential component of the head of the

chlorophyll molecule, and is also required in a number of biochemical reactions inside the cell, not as a reactant but as an activator whose function is to assist the reaction. Potassium is very important in the osmotic systems of plant cells, being present in the cytoplasm and vacuole at quite high concentrations. It is partly responsible for reducing the chemical potential of the water molecules inside the cell, thus causing the inward osmotic movement of water and the generation of the cell's hydraulic pressure. The movement of potassium into and out of a cell can therefore regulate its hydraulic pressure, and in this way the chemical plays an important role in certain plant movements, notably those of the stomatal guard cells (Chapter 9), and those of the leaves of plants such as *Albizza*, *Mimosa* and the bean *Phaseolus*, all of which show dramatic leaf movements.

The micronutrients also have many highly specific and vital functions and in their absence a plant quickly begins to look unhealthy and eventually dies. Iron is an important component of the cytochromes – compounds that are essential for electron transport in many reactions, especially in photosynthesis and respiration. Iron is also required for the synthesis of chlorophyll, which is why iron-deficient plants look pale green or yellowish. Of the remaining micronutrients, manganese, copper, zinc, molybdenium and boron are all required as components of particular enzyme systems in the cell. They act as co-factors, without which the enzymes do not work. Molybdenum and iron are essential for atmospheric nitrogen fixation. Just what boron does is not known for certain, but boron-deficient plants soon stop growing and die. Chloride is also essential for healthy growth, but again no one is quite sure why! However, while micronutrients are absolutely essential for healthy plant growth, they can quickly become toxic if present in anything more than trace amounts.

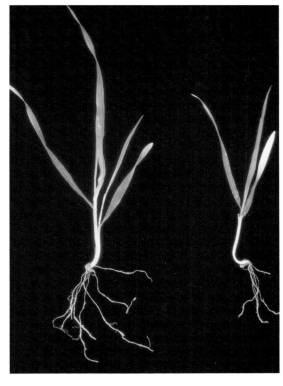

Nutrient deficiencies are often revealed by stunted growth or by a weak and sickly appearance. The spring barley seedlings on the right show typical symptoms of potassium deficiency, while the photograph below shows how seedlings of rye grass (*left pair*) and oil seed rape (*right pair*) react to normal and inadequate supplies of sulphur.

Absorbing ions from the soil solution

We must now examine just how the cells in a plant root select particular ions from the soil solution and pump them inwards across their cell membranes against a concentration gradient. The internal concentration of some ions in the cytoplasm can be up to 1,000 times greater than that in the soil solution, so quite substantial amounts of energy have to be utilized in the process.

In theory, ions could diffuse across the cell

membrane, but such an inward movement would depend upon two factors; firstly a high permeability of the membrane to the ions, and secondly the existence of a difference in the concentration of the ions on opposite sides of the membrane. In fact the plasmalemma is only slightly permeable to the water-soluble inorganic ions, and the passage of these ions through the membrane occurs only in minute channels formed in some of the large protein molecules embedded in the membrane. These proteins are called permeases. The permeability of the plasmalemma is quite different for different ions: potassium passes through most readily, while sodium and chlorine ions pass through much less readily. The passage of a particular ion through the membrane may be associated with a particular type of membrane protein. For example, it has been found that a particular protein called valinomycin can be incorporated into a membrane and, as a result, the membrane's permeability to potassium ions is enormously increased. But passive diffusion of ions into plant cells can not be a significant mechanism in ion absorption by roots because it can not account for the movement of the ions inwards from an area of lower concentration to an area of higher concentration. Diffusion only takes place *down* a concentration gradient, and in this case the concentration gradient across the cell membrane is in the wrong direction. Ions could, of course, diffuse *out* of a cell, but the relative impermeability of the membrane limits any such losses. However, if the membrane is damaged by being heated up, frozen, or treated with certain chemicals like chloroform, it becomes leaky and the solutes diffuse out rapidly.

Diffusion, then, clearly can not possibly account for the inward movement and accumulation of ions against a concentration gradient. Such an accumulation can be achieved only by the expenditure of energy, and this involves the operation of ion pumps. Ions are pumped into plant cells in two rather different ways, one of which involves an indirect kind of pumping mechanism, the other a direct pumping mechanism. The first is called an electrogenic ion transport mechanism because it involves the establishment across the plasmalemma of an electrical potential. This potential will attract or repel certain ions through the membrane because the ions are themselves electrically charged and can be either positive or negative.

The principal cause of the electrical potential across the plasmalemma is a mechanism that pumps hydrogen ions or protons (H^+) from the cytoplasm out across the membrane and into the solutuions in the cell wall and surrounding space. The removal of these positively charged particles from the inside of the cell leaves the inside negatively charged with respect to the outside, and the electro-potential difference can be as much as 200 millivolts. The existence of a strong negative charge on the inside of the plasmalemma would draw positive ions though in an inward direction and there is evidence in the case of potassium ions (K^+) that their inward movement into the cells of pea roots is largely due to this mechanism.

Each ion may have a specific ionophore channel through which it can pass, and there may, for example, in pea roots, be many more ionophore channels for potassium (K^+) ions than there are for other ions. Some plant physiologists refer to this mechanism as a passive process rather than an active one, and while this is strictly true, metabolic energy is most definitely required to pump the hydrogen (H^+) ions out of the cell to establish the electrical potential in the first place, and for this reason it is referred to here as an *indirect* ion pumping mechanism.

But movement of ions across the plasmalemma under the influence of the membrane

The red pigment in red beetroot cells does not leak out through normal healthy cell membranes. This is demonstrated by the washed cube of beetroot tissue suspended over water in the left-hand test tube. However, if the tissue is suspended over chloroform, as in the right-hand test tube, the chloroform vapour renders the cell membranes leaky and the red cell contents drip down into the liquid below.

electro-potential is not the whole story because the concentration of many ions in the cytoplasm bears no relation to what would be expected on the basis of the magnitude of the potential difference across the membrane. For the negatively charged chloride, nitrate and sulphate ions, the internal concentration is much higher than would be predicted, while for the positively charged sodium cations the concentration is much lower. These ions must obviously be transported across the membrane in some way that does not depend on the magnitude of the potential difference across it.

The direct ion pumps also appear to be associated with the proteins incorporated in the plasmalemma. Each ion may have its own pump, or at least its own channel, which recognizes it and facilitates its passage through the membrane. Roots offered a mixture of potassium (K^+) and sodium (Na^+) ions will select and pump in only the potassium (K^+) ions. Indeed, there appears to be a quite specific sodium pump which works in the other direction, pumping these ions out of the cell. Similarly, the absorption of chloride ions (Cl^-) by a root is in no way impaired even if the chloride is offered to the root in a mixed solution containing several other negatively charged ions such as fluoride, iodide, nitrate or sulphate.

That the accumulation of these ions requires metabolic energy in the form of ATP is shown by the fact that if oxygen is withdrawn from the plant, or the roots are treated with specific chemicals that stop the process of respiration and hence the generation of ATP, ion accumulation rapidly stops. Indeed, the accumulated ions even start to leak out of the cells. Our knowledge of how these ion pumps actually work is very deficient, but oat roots have an enzyme in the plasmalemma that breaks down ATP to ADP and inorganic phosphate with the release of a large amount of energy. This enzyme is called ATPase, and the energy it releases could be used to pump ions across the membrane.

The movement of ions within the root
The inorganic ions absorbed by the root cells must be able to move through the root to the central core where the major long-distance transport tissues, the xylem and phloem, are located.

The outermost part of the root consists of a mass of very long and exceptionally fine hair-like cells called root hairs, which penetrate the

soil in all directions and vastly increase the surface area of the plant over which the absorption of water and mineral nutrients can take place. Then there is a large area of more or less spherical cells called the cortex. A very special layer of cells separates the cortex from the innermost core of tissue called the stele. This separating layer is the endodermis, and its cells are different from all other cells in the root because their radial walls are impregnated with a waxy impervious layer called the Casparian strip or band. The stele consists of an outermost single layer of thin-walled cells called the pericycle, and the transport tissues, the xylem and phloem, which comprise the plumbing of the plant (Chapter 10).

The soil solution can penetrate the root quite freely – there is no cuticle or outer waxy layer as there is in a stem on leaf. The dilute solution of mineral ions can therefore pass into and along the cell walls throughout the whole of the cortex, and in doing so many ions may be absorbed by the protoplasm of the cortical cells.

Thousands upon thousands of fine root hairs cover the primary root of this sweet-corn seedling – thus providing an enormous surface area over which water and mineral ions can be absorbed.

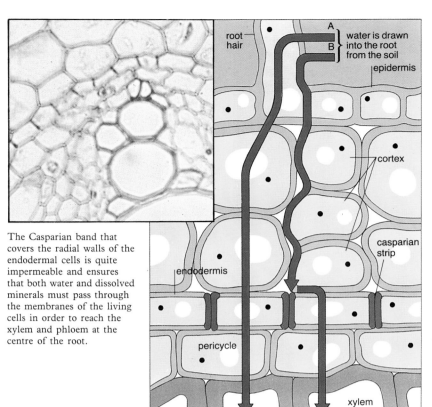

The Casparian band that covers the radial walls of the endodermal cells is quite impermeable and ensures that both water and dissolved minerals must pass through the membranes of the living cells in order to reach the xylem and phloem at the centre of the root.

Diagram labels: root hair; A; B; water is drawn into the root from the soil; epidermis; cortex; casparian strip; endodermis; pericycle; xylem

But this free movement stops completely at the endodermis because of the Casparian strip. Here, the ions can penetrate no farther unless they are taken up by the protoplasm of the endodermal cells. The endodermis is therefore the ultimate barrier: only the ions that the plant requires are selectively pumped in and eventually released into the xylem or phloem for onward and upward transport.

Nitrogen

Nitrogen, in the form of nitrate and ammonium ions, is readily accumulated by plant roots, but many soils are, or can become, very deficient in nitrogen and this leads to poor growth and the eventual death of the plant. In fact, nitrogen deficiency of the soil is probably the greatest single reason for the poor growth and low yield of crop plants in developing countries. Many plants have, however, evolved ways of thriving in nitrogen-deficient soils by utilizing either the nitrogen in the air or that present in other living organisms. The former process involves reducing the very unreactive atmospheric nitrogen gas to ammonium ions, while the latter depends on mechanisms for catching or trapping animals which are then 'digested' to release the nitrogen contained in their bodies.

How plants fix atmospheric nitrogen

The fixation of atmospheric nitrogen, that is, its reduction to ammonium ions (NH_4^+), is a process that can be brought about only by bacteria, or by a related filamentous organism called *Frankia*. No green plant has ever been found to be able to fix atmospheric nitrogen unless it has first entered into an association with one of these bacteria, and conversely some of the bacteria do not fix nitrogen unless they are associated with a green plant.

Fixation of atmospheric nitrogen takes place in a number of free-living soil bacteria, the most common being *Azotobacter*, which will operate in the presence of oxygen, and *Clostridium pasteurianum*, which will operate only in the absence of oxygen – which is frequently the case in the centres of soil particles in soils with high organic matter or humus content. These bacteria play a significant role in improving the nitrogen status of soils in many natural habitats since on their death they release into the soil the nitrogen they fixed while they were alive.

A great deal of atmospheric nitrogen fixation by bacteria is due to those species that enter into symbiotic relationships with green plants. Many plants, for example, form an association with a bacterium called *Frankia* which, at particular stages of its life, forms filamentous threads similar to fungal hyphae. *Frankia* invades the roots of many plants, notably alder, bog myrtle and various *Casuarina* species, with the result that nitrogen-fixing nodules develop.

Other associations of bacteria and green plants involve two genera of the photosynthetic cyanobacteria (formerly called the blue-green algae). These organisms, *Anabena* and *Nostoc*, grow as long chains of cells. They can fix nitrogen in the presence or absence of oxygen, but when oxygen is present only those species whose chains include large, thick-walled cells called heterocysts, fix nitrogen. The heterocysts appear to be specially modified cells in which part of the photosynthetic mechanism (Photosystem II) is missing so that no oxygen is produced in photosynthesis. The importance of keeping the oxygen concentration very low will be apparent later.

Nostoc and *Anabena* establish symbiotic relationships with a wide variety of different plants. The best-known are the lichens (Chapter 3). These are usually associations of fungi and algae, but in a number of species they are associations of a fungus and a cyanobacterium.

The colourful lichen *Rhyzocarpon geographicum* (above) grows on the surface of rocks. The 'plant' consists of an alga and a fungus living together essentially as a single organism.

Azolla is a tiny floating water fern which forms an association with the nitrogen-fixing bacterium *Anabena* to provide a continuous supply of that essential mineral.

Another well-studied association is that between the floating water fern *Azolla* and the bacterium *Anabena azollae*. This minute water fern has a horizontal floating stem with small, green, deeply lobed leaves. Filaments of *Anabena* are found associated with the apical growing point of the *Azolla* stem. Pieces of the filaments become enclosed in the lobes of the leaves as they form, and are thereby afforded protection. The way in which this happens is

that a small depression forms in the lower surface of the upper leaf lobe, and in this depression there is a prominent branched hair. The *Anabena* cells become associated with this hair, which presumably fulfils an absorption function, and eventually the original depression in the leaf surface develops into a cavity, in which the *Anabena* cells become confined. As the cavity is finally closed, the *Anabena* cells form heterocysts and the fixation of atmospheric nitrogen begins – with the result that ammonium ions are released into the cavity and taken up by the fern. *Azolla* grows widely in tropical countries, especially in the paddy fields of China where it is used to supply nitrogen to the growing rice crop. It has been estimated that floating *Azolla* with its associated *Anabena* can contain as much as 50 kilogrammes of nitrogen per hectare. As the rice grows taller, the *Azolla* is shaded and dies – releasing its nitrogen for the rice crop. Increases of rice yield of up to 20 per cent have been achieved with the proper use of *Azolla*, and under ideal conditions the plant can fix about ten kilogrammes of nitrogen per hectare, per day – completely free of charge!

By far the most important plant-bacterium association from the agricultural and economic point of view is that between *Rhizobium* and the legumes, the plant family that includes peas, beans and clovers. We now know a great deal about how the bacterium initiates nodule formation in the host root, and the biochemical processes involved in nitrogen fixation.

The first stage in the association between a growing root and the free-living *Rhizobium* bacterium in the soil is one of recognition. A large number of different bacteria live in the soil, and many are disease-causing pathogens with which the host plant must avoid forming a relationship. There must, therefore, be some mechanism by which the host plant and the *Rhizobium* cells are able to recognize one another, and this appears to be achieved by the precise arrangement of short chains of sugar molecules on the cell surfaces of both the root and the bacterium. The root secretes a protein molecule called a lectin, which has the capacity to recognize and bind precisely with a particular type of sugar molecule. The lectin is rather like a double-headed key that is able to fit into two locks – it fits both and can bind them together, providing the locks are the same. This immediately has the effect of binding only the correct bacterium to the root surface, which

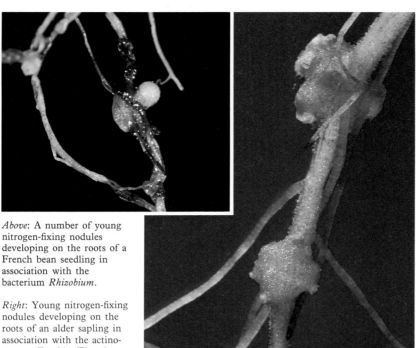

Above: A number of young nitrogen-fixing nodules developing on the roots of a French bean seedling in association with the bacterium *Rhizobium*.

Right: Young nitrogen-fixing nodules developing on the roots of an alder sapling in association with the actinomycete *Frankia*. The photograph at far right shows mature nodules on the roots of an alder.

in practice is the surface of one of the root hairs. Once this binding is complete, the root hair curls, and the *Rhizobium* cell penetrates it by stimulating the cell wall to grow inwards to form a fine tube. Within this, the bacterium cells live and divide, so increasing in number. This fine tube, the infection thread, penetrates through successive layers of cells until it reaches the inner part of the cortex where it stimulates the cells to divide. As a result of this cell division, a knob or nodule appears on the root surface. The bacteria are eventually released from the infection thread into the cytoplasm of the cells in the centre of the nodule where they enlarge and develop into curious shapes. They are then called bacteroids, and it is in this form that they have the capacity to fix atmospheric nitrogen. The root nodule thus consists of an outer region of cells not invaded by the bacteria and an inner region in which the cells contain the bacteroids. The cells in the central zone also begin to synthesize a pink substance called leg-haemoglobin (legume haemoglobin) similar to the haemoglobin in our blood, the function of which will be evident later.

Why, we may wonder, do the plant and bacterium have to enter into such a complicated arrangement in order to fix nitrogen successfully? The answer seems to be that the enzyme nitrogenase, which brings about the fixation of nitrogen, binds just as well to oxygen as it does to nitrogen. Worse still, in the presence of oxygen it binds *more* readily with oxygen, to the exclusion of nitrogen, so that no nitrogen is fixed. Only under anaerobic conditions, that is, when oxygen is excluded, can the enzyme bind with the nitrogen and fix it in the form of ammonia. The bacteroids in the centre of the root nodule are kept in an almost anaerobic atmosphere firstly by the surrounding nodule tissue and secondly by the presence of the leg-haemoglobin, which binds to, and absorbs, any oxygen molecules that might penetrate the cortical tissue. So, the nodule is a device to ensure that the *Rhizobium* is kept away from oxygen molecules so that its nitrogenase enzyme can bind freely with the available nitrogen molecules. The fixation of nitrogen by the cyanobacteria *Anabena* and *Nostoc* takes place only in the thick-walled heterocyst cells, and the reader may recall that Photosystem II of the photosynthetic mechanism is absent from these cells. Both the thick cell wall and the absence of the oxygen-generating Photosystem II appear to be associated with the need to keep oxygen away from the nitrogen-fixing enzyme nitrogenase.

One interesting feature of the association between *Rhizobium* and the leguminous plants is that it does not occur if the plant already has a plentiful supply of nitrogen in the soil in the form of nitrate ions (NO_3^-). The absence of root nodule formation under these conditions seems to be due to the fact that the recognition system is inhibited in some way, perhaps by a reduction in the production, by the root, of the critical binding and identifying factor, lectin.

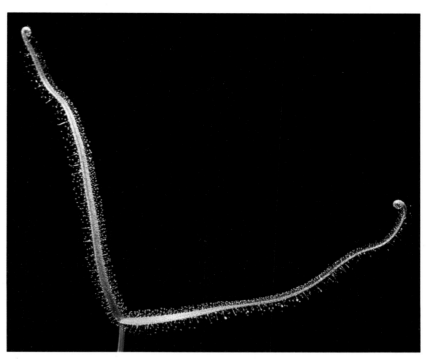

Carnivorous plants

Quite a large number of plants live in acidic habitats like bogs, where there is little or no nitrogen in the soil water, and also lack the capacity to form a symbiotic association capable of fixing atmospheric nitrogen. As an alternative, these plants have developed a taste for animals, which they catch and digest to release the nitrogen incorporated in their tissues.

Many different mechanisms have been evolved to trap insects and other small animals, and they can be divided into active and passive devices. Active devices include the extremely rapidly operating Venus fly trap. This plant has a sophisticated structure that is discussed in detail in Chapter 13 because it possesses what amounts to a nervous system, as well as a memory and the ability to count! Initially, trigger hairs are stimulated and the trap shuts. Other sensors on the trap surface then detect whether or not protein has been captured and, if so, special glands secrete enzymes which digest the animal. Another trap system which operates instantaneously is found in the aquatic plant *Utricularia*, which is also described in Chapter 13.

A slower, less dramatic, but nevertheless active system is found in the sundew, in which the leaf surface is covered in long tentacles on the ends of which are drops of a sweet but very sticky liquid. The insects stick to these liquid drops, but the tentacles then respond and curve over the insect so that it becomes firmly held both by the adhesive and by the cage of tentacles. The digestion then begins. In the butterwort, the upper surface of the leaf is covered with minute sticky glands which provide an effective fly-paper to capture insect

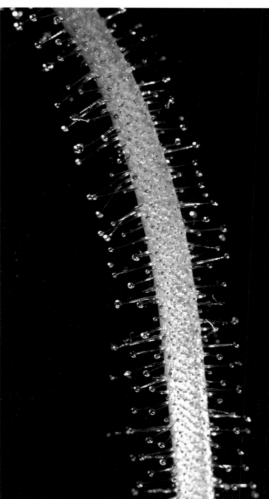

The leaves of the insect-eating sundew plant (*Drosera* sp.) are covered with tentacles, at the ends of which are sticky droplets on which insects and other small animals become trapped.

prey. Once an insect is caught, the leaf margins fold upwards and over the animal. Special enzymes are then secreted and digestion begins.

Other species employ wholly passive mechanisms, which involve no movement by the plant. These trap mechanisms usually involve the animal falling or flying into a hollow structure that has slippery, waxy sides and a narrow entrance that makes escape virtually impossible. At the bottom of these traps there is usually a pool of liquid which has a high acidity and contains enzymes that digest the prey. The pitcher plant *Nepenthes* is probably the most familiar plant in this category. The mouth of the trap is usually attractively coloured and has

been reported to secrete an attractive sugary solution. Insects alight and walk around the rim, only to discover too late that it is very slippery. They fall into the cavity, to drown in, and be digested by, the liquid at the bottom.

Another equally impressive plant is *Sarracenia*, whose leaves are modified into long trumpet-shaped pitchers, each having an attractively coloured lid which is held in the open position. Again, insects alight to inspect what to them looks like an attractive source of food, only to slip and fall into the digestive juices in the lower part of the trap. *Sarracenia* appears to have been the plant that inspired John Wyndham when he wrote *The Day of the Triffids*, and this absorbing book illustrates perhaps how terrifying these plants must be to small animals.

Mycorrhizal associations

Green plants also form another type of mutually beneficial association with the fungi, and the principal benefit of this association to the green plant is that it facilitates the uptake of mineral ions from the soil. These associations, involving the roots of plants and fungi, are called mycorrhizal associations and they are of two principal types, ectotrophic and endotrophic.

Many forest trees, particularly pine, beech and birch, form ectotrophic mycorrhizal associations with fungi. In this type of association the root becomes encased in a sheath of fungal tissue except for the growing apex. In many plants, short lateral roots become totally enclosed in the fungal sheaths. From the inner side of the sheath, fungal threads, or hyphae, penetrate the root cortex by growing between the cells in the intercellular spaces. A number of these hyphae actually penetrate the cortical cells. From the outside of the sheath the fungal hyphae penetrate far into the soil, and so clearly enlarge the volume of soil that can be tapped as a source of mineral ions. The other side of the relationship is that the fungus, being unable to carry out photosynthesis itself, gains a ready source of carbohydrate, and hence energy, from the green plant. The importance of ectotrophic mycorrhizae in the mineral nutrition of forest trees can not be overemphasized since the fungi especially exploit the mineral ions being released from the decaying leaves in the litter layer of the forest floor. Many of the toadstools found in forests are the reproductive structures of fungi involved in mycorrhizal associations.

Like those of *Nepenthes* illustrated on page 117, the pitchers of *Sarracenia* (*below*) are attractive to insects – lethally so to any insect foolish enough to attempt a landing on the slippery rim. Once inside the smooth-walled container there is no escape: the insect falls into the digestive juices at the bottom.

The beautiful toadstools of *Boletus scaber* and *Amanita muscaria*, for example, are commonly found near birch trees, a tree species with which both fungi form mycorrhizal associations.

Endotropic associations, on the other hand, do not involve the formation of a sheath of fungal tissue around the roots. These associations are found especially in orchids and they appear to be a prerequisite for really successful growth. The soil fungi penetrate the cortical cells of the orchid's root and form tight knots of hyphae. These invaded cells are often limited to the outermost three or four rows of the cortex. The fungus tissue within the innermost cortical cells appears to be digested, thus facilitating the transfer of nutrients from the fungus to the root. As with the ectotrophic associations, a large number of fungus species are involved in these associations. Some are rather specific, one fungal species being found only in a particular plant species, whereas others are quite non-specific, the same fungus forming associations with a wide range of green plant hosts. The common soil fungus *Rhizoctonia* associates with a number of orchids, as does *Armillaria mellea*.

A number of orchids are unable to carry out photosynthesis as they have no chlorophyll with which to capture the sun's energy. The plants must therefore acquire their carbohydrate nutrients (energy) from other organisms, and in these species it is the fungus that provides not only the mineral nutrients but the carbohydrates as well. Many of the fungi are able to break down cellulose in the cells walls of dead leaves and stems, and the glucose thus released can be absorbed by the fungal hyphae, transported to the orchid root, and released in the cortical cells. A few fungi are apparently able to attack and penetrate photosynthetic plants and at the same time form mycorrhizal associations with others which lack chlorophyll, in which case the fungus seems to act so as to withdraw nutrients from the green plant and transfer them to the non-photosynthetic one (Chapter 17). The exact mechanism whereby the nutrients are transferred to the root cells is not fully understood. Digestion of the hyphae may play a part, but other direct transport systems which transfer the nutrients from the fungal hyphae to the root cell may be involved. It is thought that while the digestion of the fungal threads may indeed facilitate transfer of nutrients, its main purpose is to limit the spread of the fungi to the inner layers of the cortex. Obviously, it would be self-defeating for both partners if the orchid root were to become extensively infected and damaged by the uncontrolled spread of the fungus through the whole of the root tissues.

The ways in which plants obtain their mineral nutrients are therefore many and varied. Some plants seem to manage on their own to absorb ions from the soil, but it is interesting to note that almost all roots are invaded to some extent by fungal threads. So far, however, the extent to which these threads contribute to inorganic ion absorption has proved difficult to establish.

The photographs below show the reproductive structures, or 'fruiting bodies' of two species of *Russula* which form mycorrhizal associations with the roots of trees.

CHAPTER 12

Do Plants Breathe?

Respiration is the process whereby the energy stored in sugar molecules is released and repackaged so that it can be used to drive other chemical reactions and processes such as synthesizing proteins, fixing nitrogen and transporting ions across membranes. Respiration occurs in all living cells, plant as well as animal. It involves the consumption of oxygen, and the final products when a sugar molecule is completely broken down are carbon dioxide and water. The process is very similar to combustion: if sugar is burnt in the presence of oxygen, carbon dioxide and water are produced and the energy is released as heat. Indeed, respiration has been described as the burning of sugar, in a water medium at room temperature, without the production of smoke! A major difference, however, is that in respiration a substantial proportion of the energy is conserved in other molecules so that it can do useful work within the cell; only a small proportion is lost as heat.

Many combustion reactions take place in a single step – that is, by a single chemical reaction, sometimes with explosive violence in which a huge amount of energy is released instantaneously. In such cases the final products of complete combustion, CO_2 and water, are produced directly. Such a combustion reaction takes place in the cylinder of a petrol engine when the sparking plug ignites the petrol and air mixture. The energy released is used to drive the piston down the cylinder and provide the motive force for the motor-car, although much of the energy is lost in the form of heat. But for respiration to take place by a single reaction like this in living cells would be disadvantageous for two reasons. First, the release of all the energy in the sugar molecule

at once would make it difficult to repackage in a usable form, and secondly all the carbon molecules captured in the process of photosynthesis would be lost from the plant as the CO_2 escaped into the atmosphere.

In fact, the process of respiration in plant and animal cells takes place in a large number of sequential steps, each involving a separate chemical reaction. In a number of these reactions a small fraction of the energy contained in the sugar molecule is released and stored in a temporary form. Many intermediate chemical compounds are formed before the final production of CO_2 and water. These intermediate compounds are extremely important because many enter directly into other chemical reactions and are therefore used in the synthesis of a very wide variety of chemical substances essential for the healthy growth and operation of a living cell. Many of the carbon atoms fixed in photosynthesis are therefore conserved within the plant.

Animal and plant cells respire by almost identical mechanisms. In animals, respiration is sometimes confused with breathing, but the two functions are quite distinct. Breathing is the physical process of taking in and expelling air from the lungs. Oxygen is supplied to, and CO_2 removed from, the respiring body cells via the blood stream – and the lungs provide a ventilating mechanism for adjusting the levels of these gases in the blood. Respiration is the sequence of chemical reactions that breaks down sugar, and also some other energy-storage molecules like fats, and in doing so, consumes oxygen and produces CO_2 and water.

So we return to the question of whether or not plants breathe. And the answer is that they do, in a way, because they consume oxygen

The brown/purple spadix in the centre of an *Arum* flower has such a high respiration rate that its temperature is usually several degrees higher than the surrounding air. The heat helps disperse the volatile substances whose powerful scent of rotten flesh attracts the flies that are the plant's principal pollinators.

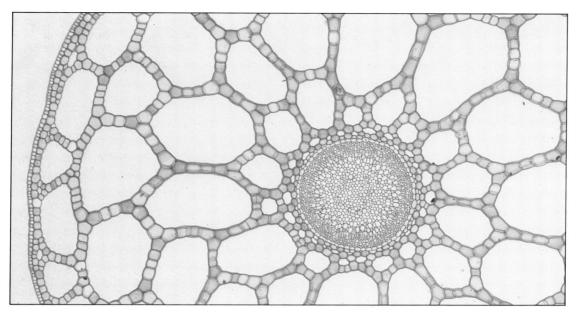

and evolve CO_2 in respiration, and these gases have to be exchanged with the surrounding atmosphere. Gas exchange in plants is, however, a passive process involving physical diffusion through pores, or through and between the surface layers of cells themselves. There are usually extensive air spaces between the cells in plant organs and it is through these that gaseous diffusion takes place inside the plant. Plants do not have an active ventilation system comparable to the breathing system in animals, but they certainly 'breathe' in the sense that they take in oxygen and emit carbon dioxide at night, when photosynthesis is not taking place. During the day, of course, this exchange is reversed in green tissues by the occurrence of photosynthesis, when carbon dioxide is absorbed and oxygen is released. Respiration does continue in plant cells during the day, but the gas exchange to which it gives rise is masked by the greater exchange arising from photosynthesis.

In Chapter 9 we discovered that in photosynthesis the solar energy is first stored temporarily in ATP molecules and in reducing power (NADPH) before these compounds are used to reduce carbon dioxide to sugar and other carbohydrates. In respiration the energy released in stages from the sugar molecule is also transferred to ATP and to reducing power molecules such as NADH and $FADH_2$, and it is the energy contained in these molecules that is then available to drive other chemical reactions in the plant cell. FAD is another molecule like NAD and NADP which has the

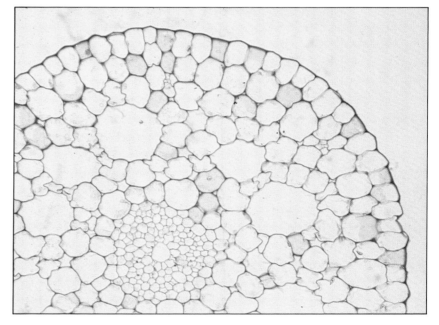

capacity to combine with hydrogen and donate it again to a reaction where a hydrogen atom is required.

The breakdown of sugar to carbon dioxide and water involves a series of chemical reactions which can be divided into two groups. Those of the first group are not in any way dependent upon oxygen. They are the reactions of a process called glycolysis, the final product of which is a substance called pyruvic acid which contains three carbon atoms. In the presence of oxygen this substance enters a second cyclical group of reactions called the Krebs cycle, which breaks it down to CO_2 and water. However, in the absence of oxygen, the pyruvic acid is dealt with by a different series of reactions altogether, and this series produces carbon dioxide and ethyl alcohol. This breakdown of sugar in the absence of oxygen is the process of fermentation, and it forms the basis of the wine-making and brewing industries in which the sources of sugar are, respectively, grapes and the endosperm of barley grains. (Spirits are produced by distillation of the fermentation mixture to concentrate the alcohol, different spirits being produced by the fermentation of different original sources of sugar. Scotch whisky is produced from barley, bourbon from sweetcorn and rum from cane sugar molasses.)

The reactions of glycolysis and fermentation take place in the cytoplasm of the cells, while those of the Krebs cycle are confined to very important, small particles called mitochondria, which are present in the cytoplasm in large numbers. There is, therefore, a physical separation of the processes involved in respiration in living cells.

To complicate matters further there is yet another, totally different, series of reactions that also leads to the breakdown of sugar and the production of pyruvic acid. These reactions operate in addition to, but in parallel with, those of glycolysis, and they are important because they involve the production of intermediate compounds with five carbon atoms, none of which occur in glycolysis. For this reason the second pathway is called the pentose phosphate shunt. We need not look at the process in detail but we should note that it is important because the five-carbon-atom intermediates are required for building the polymers incorporated into cell walls, and in the formation of molecules such as the nucleic acids.

We shall now examine some of these processes in a little more detail so that we can identify where the energy is released and how carbon dioxide is produced and oxygen utilized.

As was mentioned earlier, not all the energy stored in the sugar is released in a usable form such as ATP, or in reducing power molecules like NADH and $FADH_2$: some escapes as heat and is dissipated in the atmosphere. In plant organs with a very high rate of respiration, it is actually possible to monitor this heat production by measuring the temperature, and in some of the arum lilies this heat production is quite dramatic. The plants *Sauromatum guttatum* (voodoo lily) and *Arum maculatum* (lords-and-ladies) have flowers with a long, tongue-shaped, central spike called a spadix. This usually has the dark red or purple colour and stench of rotting meat, presumably to attract the flies upon which pollination of the flower depends. During the period of maximum odour production the temperature of the spadix can be as much as 10°C to 15°C above that of the surrounding air, and the production of CO_2, a measure of respiration rate, is very high indeed. In an *Arum* spadix at least, there can be no doubt that a significant portion of the energy stored in the sugar molecule is lost as heat. Why? It appears that the heat generated in the spadix is necessary to cause the evaporation of the extremely strong-smelling chemical compounds it contains, so that the repulsive odour of rotting flesh will attract the flies without which the plant is unable to reproduce!

Glycolysis

Reduced to its bare essentials, glycolysis involves the breakdown of sugar, which contains six carbon atoms, to two molecules of pyruvic acid, each of which contains three. Most of the energy in plant carbohydrate is stored as starch, a polymer of glucose, or as sucrose, a double sugar molecule consisting of one molecule of glucose joined to one of fructose. Before the process of glycolysis can begin, the starch must be broken down into its component glucose molecules. As we saw in Chapter 5, this can be achieved by the enzyme called α-amylase, although a number of other enzymes can also participate in the process. Sucrose is broken down into glucose and fructose by the enzyme invertase, and both glucose and fructose can then enter into the glycolysis reactions.

Before either glucose or fructose can enter into glycolysis, however, they have to be activated by the addition of a phosphate group (PO_4). This requires energy, which comes from ATP, as does the phosphate group. In the case of glucose, the resulting compound is called, quite simply, glucose phosphate. This is then converted by an enzyme to fructose phosphate which, in turn, has a second phosphate group added to form fructose diphosphate, a reactive molecule. Fructose itself can also be converted to fructose phosphate, and then to fructose diphosphate by ATP and the appropriate enzymes. A very important enzyme called aldolase then splits the fructose diphosphate molecule into two slightly different compounds, each containing three carbon atoms and one phosphate group. These compounds exist in equilibrium, and are readily interconverted, but only one, called phospho-glyceraldehyde, is fed into the next set of reactions. These reactions lead to the production of pyruvic acid, another important molecule containing three carbon atoms.

Two molecules of ATP are synthesized for each molecule of phospho-glyceraldehyde converted to pyruvic acid, and it must be remembered that two molecules of phospho-glyceraldehyde are produced from each sugar molecule. So although two ATP molecules were used to activate each sugar molecule, *four* molecules of ATP are produced by the time this molecule has been broken down to two molecules of pyruvic acid. This net gain of two ATP molecules is an example of the gradual release of the energy stored in the sugar molecules and the way in which it is made available

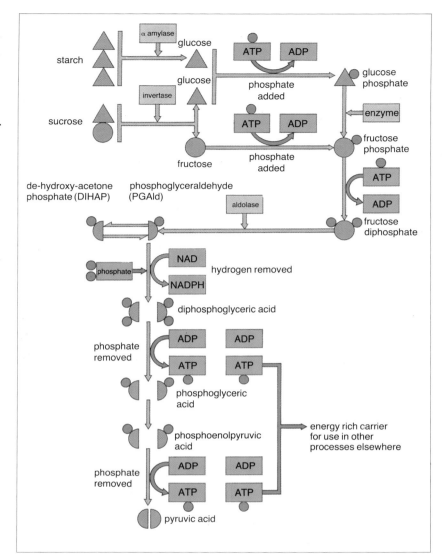

in the usable form of ATP. In addition to this, two molecules of reducing power in the form of NADH are produced each time a molecule of sugar is broken down into two molecules of pyruvic acid. The energy in these two molecules of NADH can be used in the mitochondria to produce a total of six molecules of ATP during their re-oxidation to NAD^+. Their reducing power can also be used directly in other reactions where electrons are required. The continuation of glycolysis depends on a supply of NAD^+ and this is provided by the mitochondria re-oxidizing the NADH. If the cell is deprived of oxygen the mitochondria obviously cannot carry out this process and consequently there is a build up of NADH in the cell. This NADH can, however, be re-oxidized to a very limited extent in the absence of oxygen by the process of fermentation, in

The diagram above shows in some detail how molecules of the sugars glucose and fructose are broken down into pyruvic acid by the process of glycolysis. The glucose and fructose are derived initially from the breakdown of starch and sucrose by enzymes. They are then 'activated' by the addition of phosphate groups, which are supplied by ATP. They then enter the main glycolysis reactions which result in two molecules of ATP and two of pyruvic acid being released for each molecule of sugar broken down.

which the pyruvic acid is converted to alcohol (a molecule containing two carbon atoms) and carbon dioxide.

Under anaerobic conditions, when the mitochondrial oxidizing systems cannot operate, the pyruvic acid is converted first to a two-carbon compound called acetaldehyde with the loss of a molecule of CO_2. The acetaldehyde is then reduced to ethyl alcohol by the NADH, thus releasing the oxidized form NAD^+ which can be recycled into the glycolysis reaction in which it becomes again reduced. In some plant tissues, lactic acid is produced instead of alcohol. In animal cells that are short of oxygen, lactic acid rather than alcohol is produced – and that is one of the causes of the muscular cramp and stiffness that often follow violent exercise. If animal cells produced ethyl alcohol when oxygen was in short supply, the inebriating consequences hardly bear thinking about!

Under anaerobic conditions, then, plant cells produce only two molecules of ATP per molecule of sugar respired, but in the presence of oxygen, 36 molecules of ATP or usable energy are produced per molecule of glucose consumed. The release of available energy is thus dramatically reduced when oxygen is in short supply. Plant organs, especially roots and underground stems, quite often become deprived of oxygen when the soil becomes waterlogged, and the resulting reduction in energy release from sugar molecules drastically reduces the rates of growth and other vital processes, such as the uptake of nutrients from the soil and the synthesis of cellulose. Whilst the roots of many plants can withstand waterlogging for some days, others can not. If roots die due to a failed energy supply and the accumulation of alcohol, then the whole plant will die. On the other hand, some plants thrive in the mud at the bottom of ponds and can tolerate a persistent shortage of oxygen. Rice, for example, needs to be waterlogged in its early stages of growth, although not when the plant is mature.

In a plentiful supply of oxygen the process of glycolysis is exactly the same as in the absence of oxygen, but the pyruvic acid produced is not converted to alcohol and CO_2: instead it is converted completely to CO_2 and water by being fed into the Krebs cycle. So let us see what is involved in this alternative series of reactions which is so effective in producing 34 additional ATP molecules compared with the zero addition produced in fermentation.

The Krebs cycle

The reactions of the Krebs cycle take place in the sub-cellular particles called mitochondria, which means that the machinery for breaking down pyruvic acid to carbon dioxide and water with the production of all the additional ATP molecules is separated from the reactions of glycolysis, which take place in the body of the cytoplasm itself.

The mitochondria are usually sausage-shaped bodies located in the cytoplasm. They have a rather complicated internal structure of membranes reminiscent of chloroplasts. There are two layers of membrane surrounding a mitochondrium, the outer one being perfectly smooth while the inner one has a large number of inwardly-directed folds which have the effect of dividing the space inside into a series of compartments. The inwardly-projecting folds of the inner membrane are called cristae, and they are important in that they contain the enzymes responsible for bringing about the electron transport process associated with the uptake of oxygen, without which the Krebs cycle could not operate. (The situation is similar to that in which the electron transport involved in Photosystems I and II, and the *release* of oxygen in photosynthesis, takes place in the inner thylakoid membrane of the chloroplast.)

The Krebs cycle itself comprises a series of reactions involving organic acids that have four, five or six carbon atoms, and is really quite easy to follow once it is appreciated that in feeding in one pyruvic acid molecule containing

Under an electron microscope, a mitochondrion can be seen to have a complex double membrane – the outer one relatively smooth, the inner one forming a large number of inwardly projecting folds called cristae. The folds contain enzymes essential to the operation of the Krebs cycle.

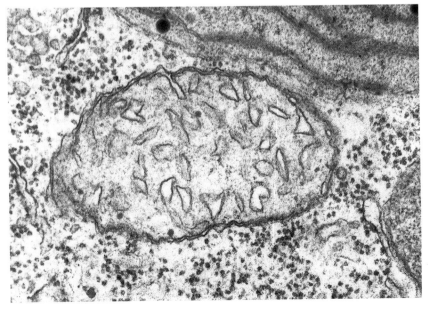

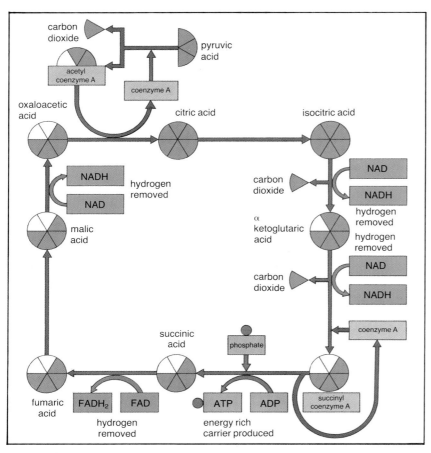

The details of the Krebs cycle shown above reveal how the three atoms of carbon in pyruvic acid are released as three molecules of CO_2, and how, in the process, reduced (that is, hydrogen-containing) forms of NADH and $FADH_2$ are produced. Each turn of the cycle also produces one molecule of energy-rich ATP, which can then be used elsewhere. Each of the purple segments represents one carbon atom.

three carbon atoms, these three carbon atoms have to be lost as three molecules of the gas CO_2 in the course of one complete turn of the cycle.

In fact, only two carbon atoms are actually fed into the cycle because pyruvic acid first reacts with a sulphur-containing compound called coenzyme A. In this reaction one molecule of CO_2 is evolved from the pyruvic acid and the remaining two carbon atoms join with coenzyme A to form *acetyl* coenzyme A. Two hydrogen atoms have to be removed in this reaction so that a molecule of NAD^+ is reduced to NADH.

The acetyl coenzyme A now donates its two-carbon portion to an organic acid called oxaloacetic acid having four carbon atoms to give a six-carbon acid called citric acid. After this reaction, there follow seven more, which lead to the regeneration of oxaloacetic acid with four carbon atoms, and during this cycle two carbon atoms are lost as two molecules of CO_2. The steps in this reaction are shown in the summary diagram, but the main details of importance to the reader are the points at which CO_2 is evolved, and the reactions that

are dependent on a supply of the oxidized form of the coenzymes NAD^+ and FAD, which have to be available to accept the electrons and hydrogen ions released in the reactions. Since two molecules of pyruvic acid are produced for each sugar molecule that enters into glycolysis, two turns of the Krebs cycle are required to release as carbon dioxide the six carbon atoms in each sugar molecule. It will be noticed that the two electron or hydrogen acceptors used in Krebs cycle are NAD^+ and FAD, not $NADP^+$ which plays such a prominent role in electron transport in the chloroplast in photosynthesis.

The oxygen requirement

Readers may be wondering why no mention has yet been made of the involvement of oxygen in the processes. Oxygen, after all, is critical for the support of life; all animals and most plants die in its absence. All we have seen, so far, is a series of chemical reactions that can lead, in air, to the progressive breakdown of a sugar molecule containing six carbon atoms and the loss of these carbon atoms as six molecules of the gas carbon dioxide. No reaction in either glycolysis or the Krebs cycle involves oxygen. So why is oxygen essential for respiration and life? The answer to this question is quite simply that the Krebs cycle only functions when it is supplied continuously with the oxidized forms of the coenzymes NAD^+ and FAD. From the accompanying diagram it will be clear that during each cycle three molecules of NAD^+ are reduced to NADH, and one molecule of FAD to $FADH_2$, and if these reduced, or hydrogen-containing, molecules of NADH and $FADH_2$ are not re-oxidized to regenerate NAD^+ and FAD, then the Krebs cycle will stop and the process of fermentation will begin. So the involvement of oxygen in respiration concerns the re-oxidation of the reduced coenzymes NADH and $FADH_2$, that is, the removal of their hydrogen atoms, as a result of which water is produced.

This oxidation reaction does not take place in one step but rather by a series of reactions, rather like those we saw in the process of electron transport in photosynthesis, except that different electron transporting molecules are involved. The chain of molecules involved in this electron transport process all appear to be located in the inner membrane forming the cristae which project into the innermost part of the mitochondrium. What is of the greatest importance is that in the oxidation of $FADH_2$

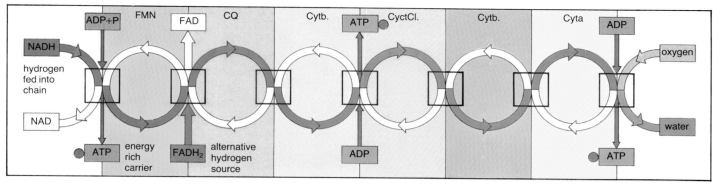

The Krebs cycle on the facing page can only keep going if it is supplied continuously with NAD and FAD. These leave the cycle in a reduced state – that is, each one carries away a hydrogen atom. To be used again to fuel the cycle, each of these molecules must lose its hydrogen atom, and it does so in a chain reaction, illustrated above. Molecules of NADH and FADH$_2$ pass their H atoms into the chain where they are finally combined with atmospheric oxygen to form molecules of water. However, along the chain energy-rich ATP molecules are released, and it is this process that makes available for other uses much of the energy released when a molecule of sugar is broken down into carbon dioxide and water.

and NADH in this electron transport chain, the energy released is captured and held in the form of ATP which is synthesized from ADP and phosphate. The energy-rich ATP molecules are then able to release their energy wherever it is required.

The process of ATP synthesis during the oxidation of NADH and FADH$_2$ is called oxidative phosphorylation and it takes place only in mitochondria in the presence of oxygen. It is the reactions of oxidative phosphorylation that give rise to most of the 36 ATP molecules that are produced when a molecule of sugar is completely broken down to carbon dioxide and water.

Details of the chemical substances involved in the electron transport chain need not concern us here, except to say that a number of them are compounds called cytochromes which contain iron. All the chemicals in the chain can exist in an oxidized and reduced form, each changing to the other form as it oxidizes its predecessor and is, in turn, oxidized by its successor in the chain. The last cytochrome in the chain is oxidized by molecular oxygen from the air which, in gaining the hydrogen atoms, is itself reduced to water. The products of aerobic respiration are, therefore, carbon dioxide and water.

The way in which ATP synthesis is driven by these oxidation-reduction reactions in the mitochondria appears to be similar to that which operates in photosynthesis. The molecules involved in the reactions seem to be asymmetrically arranged on the inner membranes of the mitochondria. The oxidation and reduction of some of the chemical substances in the reaction chain involve the transfer of hydrogen ions, as well as electrons, across the inner membrane in an outward direction so that the outer compartment becomes more acid than the inside. These hydrogen ions diffuse back along the concentration gradient into the

inner compartment of the mitochondria through the membrane of the cristae at the site of specific chemical structures similar in function to the coupling factor which occurs in the thylakoid membrane of the chloroplast. Passage of hydrogen ions through this structure leads to the synthesis of ATP from ADP and phosphate.

So respiration is an extremely important series of reactions which has two principal functions. First, it releases the energy stored in sugar molecules and makes it available in a readily usable form, ATP, which can participate as an energy source in a very large number of biochemical reactions. Second, it provides a large number of intermediate chemical compounds of widely differing structure which act as the starting points for the synthesis of a multitude of complex chemicals necessary for the growth and development of a healthy plant.

How efficient is the release of the energy from sugar and its repackaging into the widely usable form of ATP? In air, the total breakdown of one molecule of sugar to CO_2 and H_2O results in the transfer of about 40 per cent of the energy to the ATP molecules. The rest is lost as heat. In the absence of oxygen only about two per cent of the energy in the sugar molecule is transferred to ATP!

In Chapter 9 we likened photosynthesis to an energy packaging and repackaging mechanism in which the sun's energy was captured and stored first in ATP molecules, and then repackaged for transport or longer-term storage in the form of carbohydrates, such as glucose, sucrose and starch. We should now think of respiration in similar terms; that is, as a process which unpacks the energy from the carbohydrate molecules and repackages it into a much more readily usable form, namely ATP molecules. However, the price of doing this is the loss of more than half of the original energy stored in the sugar molecule!

CHAPTER 13

Sensitive and Nervous Plants

Very rapid responses to external stimuli are common-place in the animal kingdom and are of obvious importance in survival. The rapidity of these responses depends upon the highly developed sensing organs and complex nervous systems which animals have developed in the course of evolution. A number of plants, too, appear to have developed sophisticated 'nervous systems' that enable them to transmit information quickly from one part of the plant to another and to react almost instantaneously to various stimuli. In some plants the reaction is used to trap insects and other prey, while in others a swift response may ensure that pollen is firmly planted on the body of a visiting bee or fly, or that tender foliage is pulled out of reach of a browsing animal.

Incredible though it may seem, the development of a nervous system has provided some plants with the ability to count, and with a memory as well. Admittedly the plant can distinguish only between zero, one and two, but even the most powerful computer in the world is based on a mechanism that can only tell the difference between zero and one! The plant's memory is also short, retaining information for only 30 to 40 seconds, yet it is a genuine memory storing very specific information.

To justify these seemingly outrageous statements we need look no farther than the many familiar plant species that have the ability to move quickly in response to a stimulus. Among the best-known are the sensitive plant (*Mimosa pudica*) and the Venus fly trap (*Dionaea muscipula*). Other good examples are provided by the telegraph plant (*Desmodium gyrans*) and by the male and female fertile flower parts of *Mimulus, Berberis, Stylidium* and several members of the orchid family.

If the extreme tip of a mimosa leaflet is squeezed gently or warmed momentarily with the flame of a match, the leaflet will immediately collapse or close up alongside the leaf stalk. Within seconds a wave of motion passes along the leaf stalk as successive pairs of leaflets collapse into a closed position. A beetle or fly may blunder about in the inviting interior of a Venus fly trap for several seconds, apparently quite safe. Then suddenly the trap is sprung. The two parts of the leaf close so quickly that only the most agile or fortunate insect is able to makes its escape.

The fact that these reactions are triggered by touch reveals immediately that the plant must have some sensory structure or structures capable of detecting physical contact. The movements themselves are brought about by dramatic changes in cells in specialized hinge regions at the base of each mimosa leaf, and along the centre of the fly trap. We will return to these mechanisms later. What is of immediate interest is the fact that the sensors are located as much as 15mm away from the motor cells that cause the movement; and since the leaf and trap movements can be completed within half a second of the plant being touched, it follows that transmission of information from the sensor cells to the motor cells must be very swift indeed. Such rapid transmission of a stimulus is most likely to be achieved by means of electrical impulses, as it is in the animal nervous system. We are faced, then, with the question of whether or not plants transmit information by means of electrical potentials; and if they do, whether this means that they really possess a nervous system.

The moist red patches on the lobes of a Venus fly trap are an enticing sight to any approaching fly, but the six trigger hairs inside the trap could spell disaster for the unwary visitor.

The Venus fly trap

The sensory system of the Venus fly trap consists of six hairs, three grouped in the centre of each of the fleshy trap-lobes. The hairs are complex structures but essentially they are transducers; that is, they are able to convert a mechanical (physical) stimulus into an electrical one. The long, stiff hair functions as a lever, and when it is pushed sideways considerable pressure is exerted on the large cells clustered around its base. These cells are highly specialized, and when they are squeezed beyond a certain point their electrical properties undergo an instantaneous change.

The system works like this. In their normal unsqueezed state the cells pump electrically charged ions outwards across the cell membrane so that an electrical potential, or voltage difference, is established between the inside of the cell and the outside. In the fly trap the potential has a steady value of about 150mV, the inside of the cell being negatively charged with respect to the outside. This steady electrical state is called a 'resting potential'. When movement of the trigger hair puts a basal cell under pressure, a point is reached at which the permeability of the cell membrane undergoes a sudden and dramatic change. All the ions

that were pumped out of the cell to create the resting potential rush back in again, and the resting potential instantaneously drops to zero, or to very near zero. This sudden drop, or discharge, is called an 'action potential' and it is the key to the plant's communication system. It is, however, a very short-lived phenomenon, for as soon as the potential is lost the cell membrane returns to normal, the charged ions are pumped out again, and within about a second the resting potential is restored.

Having thus created a pulse of electrical energy – the action potential – the plant must now transmit this signal through its tissues for a distance of up to 15mm in order to trigger the trap-closing action of its motor cells. This is achieved by a kind of chain reaction, a sequence of action potentials discharging in cell after cell in rapid succession, each one triggered by the one before. Finally an action potential is triggered in the big motor cells, but here the response is even more dramatic. In addition to losing its electrical potential, the motor cell membrane suddenly becomes completely permeable. The solution inside the cell, which has been pumped up to high hydraulic pressure (Chapter 6), floods out into the cell walls and the spaces between the cells.

A Venus fly trap in the set position (*below*) and in the sprung position (*below right*), after the trigger hairs were touched, twice, by the unfortunate fly whose wing can be seen protruding from the trap margin.

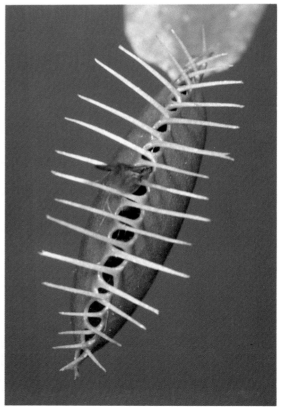

When the trigger hair is undisturbed, there is a negative electrical charge (resting potential) of about 150mV across the membranes around the sensory cells. When the cells are squeezed (at the point arrowed) the charge is instantaneously lost, but it is regained after a few seconds. This sudden discharge is called the action potential.

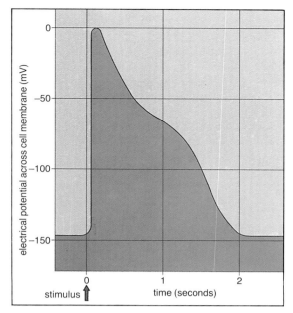

The trigger hair of a Venus fly trap consists of a long stiff lever, at the base of which are a number of large cells (shown here in red). When squeezed by the hair being pushed to one side, these cells release a pulse of electricity, the action potential, which sets off the trap-closing mechanism.

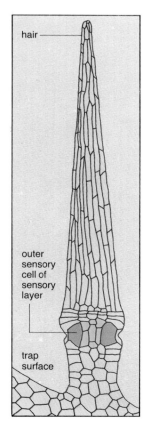

hair

outer sensory cell of sensory layer

trap surface

All hydraulic pressure is lost – and the cell collapses like a punctured football.

The rapid closure of the Venus fly trap appears to be due entirely to its springy structure. It is, quite literally, a spring trap, like an old-fashioned rat trap. In the 'normal' or 'relaxed' state the trap would lie closed. It is opened into the 'set' position by the expansion of the big motor cells as they are pumped up to high hydraulic pressure by osmosis. For as long as the pressure in these cells is maintained, the trap will remain open. But the instant pressure is lost, the trap snaps shut – usually with sufficient speed to ensnare the victim.

When the trap is sprung, it closes swiftly until the finger-like projections on the lobe margins interlock to form a cage. There is then a second, much slower, phase of closure which takes place only if a worthwhile catch has been made; and 'worthwhile', to a Venus fly trap, means protein. After the first closure, the fly trap 'tastes' whatever it has caught by means of sensing glands on the surface of its lobes. If it is a leaf fragment, a soil particle, or a bit of paper poked in by an inquisitive human, the plant wastes no further time or effort on it. Within a few hours the motor cells are pumped back up to full pressure, the lobes are opened wide, and the trap is re-set for another attempt. But if the trap contains protein, in the form of an insect, the trap closes fully. The two lobes press firmly against the victim, tiny glands on the surface of the lobes secrete enzymes to digest the food, and the trap remains closed. It may open again once the nutrients have been

absorbed, but will not function again as a trap.

One or two simple experiments are enough to show that the Venus fly trap is a very sophisticated plant indeed, and that it possesses both the ability to count, and a memory!

Using a fine glass filament held in a mechanical manipulator, it is possible to give one of the trigger hairs a single touch. Rather surprisingly, nothing happens, even though the hair may be bent right over. However, if the hair is touched a second time, the trap shuts instantly. It is clear, then, that the plant can distinguish between one touch and two; that is, it can count. In order to do this it must have a memory. The first assumption might be that information about the first touch is stored in the hair itself, but this is not so. The second touch can be given to any one of the other five hairs and the trap will respond just as promptly. The memory of the first touch may be stored in the motor cells themselves, so that arrival of a second action potential, from whatever source, will trigger the collapse of the cells. This part of the story has yet to be fully unravelled.

Although the nature and location of the memory system has not yet been established, it is possible to determine experimentally just how good a memory the fly trap has. Traps are stimulated artificially, and in each test the time interval between the first touch and the second is increased. If the second touch is given within about 40 seconds of the first, the trap will close, but if the interval is longer there is no response – the trap has by that time forgotten the first stimulus.

In human terms a retention time of up to 40 seconds is hardly impressive, but it is perfectly suited to the fly trap's particular needs. A longer memory would, in fact, be a distinct disadvantage. If the trap could be triggered by a single touch, then every random raindrop or piece of wind-blown debris would set off a useless closure. A single accidental stimulus like this must also be forgotten fairly quickly, otherwise a second random touch, some time later, would also cause an unproductive closure. By having a memory of 30 to 40 seconds such wasted trap movements are kept to a minimum. And since a fly or other insect crawling about inside the trap is almost certain to hit the sensory hairs twice within the plant's half-minute memory span, it appears that evolution has 'tuned' the plant's capabilities very precisely indeed to its environment and needs.

Snares, pit-falls and 'suction' traps

The many different carnivorous plants that capture insects, spiders, worms and caterpillars have evolved a wide variety of mechanisms with which to ensnare their victims. Some plants combine stickiness with slow movement. The butterwort, for example, has leaves with a sticky surface and operates on the 'flypaper' principle. An insect contacting the surface becomes trapped by the adhesive, after which the margins of the leaves slowly bend upwards and over the insect, which is digested where it lies. In the sundew, shiny beads of attractive-looking adhesive are located at the ends of tentacles projecting from the surface of the leaf. Once an insect becomes stuck, the tentacles slowly curl over, completely enveloping the victim.

Other plants have entirely passive trapping mechanisms in which no active movement is involved. The pitcher plant, for instance, has been discussed in Chapter 11.

Another very active, indeed explosively active, mechanism occurs in the common aquatic plant bladderwort. The much-divided leaves bear many tiny transparent bladders, each of which has an inward-opening trapdoor ringed by sensory hairs. When the trap is in the set position it is tightly contracted, but when a small aquatic organism swims along and touches one of the sensory hairs the bladder

The leaves of the butterwort (*Penguicula*) have extremely sticky upper surfaces, and to aid the digestion of trapped insects, the edges of the leaves curl up and over the victim.

suddenly expands. The trapdoor swings open under the water pressure and water is sucked into the bladder – sweeping the organism inside. Within 300 milliseconds of the stimulus the door is closed again. The volume of the bladder is gradually reduced to its former size,

The bladderwort (*Utricularia, below left*) captures small aquatic animals by sucking them violently into the tiny bladders on its stems. The action is triggered when an animal touches one of the sensory hairs at the mouth of the bladder (*below*).

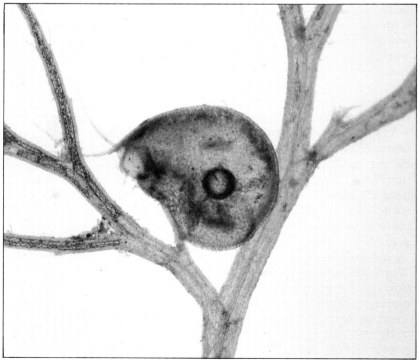

but for the organism inside there is no escape. Often a number of minute organisms can be seen inside these quick-firing bladder traps.

Exactly *how* the bladderwort trap operates is not yet known. It is a very small structure and one not easily studied in detail. However, it is known that only a single touch stimulus is needed to activate the trap, and that the re-setting mechanism involves the pumping of water and mineral ions from the interior of the bladder to the outside.

Solving the nutrient deficit

All this leads to the inevitable question, '*Why* do plants catch animal prey?', and the answer is that just like animals, plants require a variety of minerals in their diet. Most plants obtain the nitrogen they need from the soil – by absorbing ammonium and nitrate salts through their root systems (Chapter 11). However, in some habitats nitrogen in any form is almost completely absent from the soil, and it is no coincidence that it is in these environments that carnivorous plants are most commonly found. Pitcher plants, for example, are widely distributed through the tropical rainforests of Southeast Asia where soils are typically thin and poor in nutrients. Some species of butterwort are found in the most nitrogen-deficient bogs of Scotland and Ireland, while others manage to live in the dry stony soils of the high Andes.

By capturing and digesting insects and other small animals, carnivorous plants are able to secure an alternative source of nitrogen. The process of digestion releases nitrogen-rich amino acids from the animal protein of the victim's body, and these are then absorbed and used to build up new proteins in the body of the plant. In the same way, we obtain nitrogen-rich amino acids from the meat we eat.

Any doubts about the importance of this source of nitrogen to a plant such as the Venus fly trap can be expelled by performing a very simple experiment. Purchase a Venus fly trap from a garden centre and keep it in a warm room inside a plastic bag or under a glass cover to maintain the high humidity the plant requires – but feed it only on water. The plant should be given no other form of nutrient. When the plant begins to look 'sick' place a tiny sliver of fillet steak on one of the traps, moving it slightly to ensure that the trap closes. Then sit back and watch. Over the next few weeks, with no further attention, the plant will recover and flourish.

The sensitive Mimosa

The sensory mechanism of the mimosa plant is not quite as clearly defined as that of the Venus fly trap but its reactions to touch are every bit as spectacular. Very little stimulus is needed to set the plant in motion. Gently squeezing the tip of one leaflet will set off a chain reaction. One after the other, the tiny leaflets collapse alongside the leaf stalks, and even the stalks themselves eventually droop into a relaxed position. In the space of just a few seconds the entire plant appears to wilt.

Using a stop-watch it is possible to time the stimulus as it passes down the leaf-stalk from one leaflet to the next, and such experiments give a transmission speed of about ten millimetres a second. If fine electrodes are inserted at intervals along the leaf-stalk it is also possible to measure the speed at which the action potential is passed along inside the stalk. The two are found to be exactly the same, leaving no doubt at all that electrical impulses are the means by which mimosa is able to pass information along its shoots – in some cases for

The 'sensitive plant' *Mimosa pudica* can collapse all its leaves in just one-tenth of a second. Viewed from above, the undisturbed plant has its foliage spread wide to catch the sun's rays. The lower photograph was taken a few seconds later, after the pot was given a sharp tap.

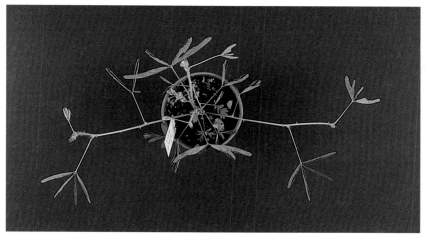

If the two end leaflets on four adjacent fronds of an undisturbed *Mimosa* plant are gently squeezed, in sequence, at intervals of a few seconds, the gradual spread of the stimulus down the leafstalks can be observed. In this 'before-and-after' pair, the lower photograph was taken just after the bottom left frond was touched. Already the first frond is fully closed while the other two are at intermediate stages.

In the Australian plant *Stylidium*, the stamen and style are fused into a structure called a column, which is normally hinged back, held under tension. When released by an insect, the column swings violently inwards, hitting the insect and covering it with pollen. At the same time, pollen already on the insect's back is picked up by the stigma. The whole exchange takes place in a split second.

quite considerable distances. When an action potential reaches the pressurized motor cells of the hinge region, or pulvinus, it causes an immediate loss of hydraulic pressure so that the leaf, or leaf-stalk, collapses. After about twenty minutes the pulvinus cells are pumped back up to full pressure, the leaflets open out again and the leaf-stalks resume their normal more-or-less horizontal positions.

In neither the Venus fly trap nor mimosa is there any anatomical evidence of a specialized conducting tissue for these electrical stimuli. The plants have nothing analogous to the highly developed nerve-fibre network of the animal body. It seems that in the Venus fly trap all the living cells of the trap lobes are equally capable of producing an action potential in response to the discharge of an adjacent cell. In the leaf stalk of mimosa there is some evidence that very elongated cells near the centre, and also near the periphery, may be specially associated with the transmission of electrical stimuli but there is, as yet, no proof that these cells are specialized for that purpose.

Sexual trickery in plants

Nowhere in the plant kingdom is botanical engineering more delicate, more precise or more ingenious than in the vital matter of ensuring that flowers are pollinated. Trapdoors, pit-falls, levers and springs are all used to gather pollen from the bodies of visiting insects, and to send each unwitting helper on its way liberally dusted with the plant's own pollen.

Rapid movement occurs, for example, in the female stigma flaps of *Mimulus* flowers. A pollen-laden insect entering the flower will stimulate the stigma and cause the flap to close swiftly but gently against its body. The result is that pollen is transferred to the inner surface of the flap. In the same way, the male stamens of *Berberis* flowers are sensitive to touch on their inner surfaces. When an insect makes contact the stamen instantly pivots inwards, thus ensuring that the pollen-laden anther brushes against the insect's body.

One very well studied case of movement in the sexual parts of flowers occurs in the Australian plant *Stylidium*. Here, the plant's stamen and style are fused together into a stout structure called a column, which is normally held back, under tension, by the swelling of the big motor cells in the hinge zone at its base. As soon as a visiting insect disturbs the

The hammer orchid (*Drakaea*) of Australia is even rougher than *Stylidium* in its treatment of insect visitors. Part of the flower mimics a female wasp, and when a male wasp arrives and tries to mate with it, his actions release a spring-loaded lever which catapults him head-first into the flower, to emerge somewhat shaken – and with pollen sacs firmly attached to his back.

column slightly, the whole structure swings over towards the rest of the flower – virtually clubbing the insect with the end of the column and ensuring that the bewildered creature departs with a liberal coating of pollen. In the same violent movement, the plant's style picks up pollen from the insect's body, so ensuring its own pollination. This remarkable two-way exchange of vital genetic material is effected in just 10 to 20 milliseconds!

An even more astonishing pollination mechanism – and probably the most traumatic of all for the insect involved – is found in the hammer orchid which grows in the dry grasslands of South Australia. After the passage of one of the frequent bush fires that occur in that region the plant sends up a very long stalk, on the tip of which there develops a single flower. Like all orchids, the flower is an elaborate structure and in this particular species it is modified to resemble the body of the female of a particular species of wasp. The wasps are unusual in that they live in the soil, and only the males can fly. In order to mate, the females must climb to the top of a tall plant stem, release their mating scent and await the arrival of a male. But the male wasps always appear about two weeks earlier than the females – and the orchid has evolved to take advantage of this time difference. The flowers open just as the male wasps appear, and they emit a chemical odour very similar to that of the female wasp. Driven by the mating instinct the male wasps land on, and attempt to copulate with, these cunningly-fashioned 'decoy' females. Unfortunately for the male, his amorous activities trigger the plant's action mechanism which causes the specialized lobe of the flower to respond with almost explosive force. The flower lobe has a spring-loaded 'elbow' joint which operates in much the same way as the *Stylidium* column – but much more violently. The unfortunate male wasp is hurled head-first into the orchid flower, turning a half-somersault in the air to crash, head downwards, with his back against the orchid's stigma and pollinia.

When the wasp recovers, he struggles to climb out of the flower and in doing so acquires two pollen-laden sacs, the pollinia, which become firmly attached to the back of his head or to his back. (Many tropical and subtropical wasps in museum displays can be seen to have these conspicuous packages stuck to their bodies). When the wasp visits other flowers, the procedure is repeated and so the pollen from the pollinia is distributed to the female stigmas of other flowers, while more pollinia may also be picked up.

This rather exciting period in the wasp's life comes to an end when the real female wasps emerge from the soil and climb high into the grassy vegetation. No matter how impressive the orchid mimics may be, they are no competition for the real thing. The males turn their attention exclusively to the real female wasps, but the trickery has worked; enough orchids will have been pollinated to ensure the next generation. Just *how* the instantaneous reaction is set off in the hammer orchid has not yet been established, but there can be no better example of the exploitation of animals by plants than this gigantic botanical confidence trick!

CHAPTER 14

Plant Clocks – The Measurement of Time

The ability to measure time is not something we usually associate with living organisms, but there is no doubt that both plants and animals have time-measuring mechanisms and can therefore organize their behaviour on a temporal basis. These time-measuring mechanisms are called biological clocks.

Two rather different lines of research led to the realization that plants can measure the passage of time. The first was the discovery that several aspects of plant behaviour, such as the movement of leaves and the opening and closing of flowers, continue in a rhythmic manner even when a plant is totally isolated from changes in environmental factors such as light intensity and temperature. Of course, in the natural environment these and other factors show a cyclical variation between day and night, and it is perhaps not surprising that plants normally respond by showing a 24-hour rhythm in their behaviour. But the fact that this rhythmic behaviour continues, often for several weeks, when plants are transferred to a controlled environment where the temperature, light intensity, pressure and humidity are held quite constant, means that there must be, within the plant, a mechanism that generates the rhythmic behaviour. The period of such rhythms is usually about 24 hours, and this has led to them being called circadian rhythms, from the latin words *circa* – 'about' and *dies* – 'a day'.

The second line of research was concerned with the onset of flowering. In a great many plants, though certainly not all, flowering occurs only at a certain time of the year, and the onset of flowering in a particular species is often timed with uncanny accuracy to occur in a particular week of the year, despite quite large fluctuations in temperature and rainfall from one year to the next. The only environmental factor that can give a living organism precise information about the time of year is, of course, the length of the day and night. But this signal can only be used if the plant has the means to measure it! Possession of a biological clock is therefore essential for the measurement of the length of the day or night.

It is now well established that plants measure the length of the night. Just how we think they do this will be described in Chapter 15, but it is an intriguing problem because darkness is in reality the *absence* of visible radiation. What plants actually do is to measure the interval that elapses between the reception of two successive light signals, dusk on one day and dawn on the following day.

The responses of plants, and indeed animals, to the lengths of the day or night are called photoperiodic responses. The phenomenon was discovered during studies of flower initiation in plants in 1920, but it is now known to have an important controlling role in many aspects of the development and behaviour of plants and animals. In plants, for example, winter bud formation, leaf shedding, the formation of underground storage organs, leaf growth and pigment synthesis may all be controlled by night length, while in animals, the onset of migration and ovulation in birds, and of ovulation in sheep, are all under photoperiodic control. However, it appears that in animals it

A view from the bank of the River Almond in Perthshire, Scotland, in mid-summer (*top left*) and mid-winter (*top right*) shows the dramatic effect on foliage that accompanies the changing of the seasons.

The rich autumnal colours of this deciduous woodland in northern Ontario, Canada, are the result of senescence, or old age, of the leaves. This characteristic seasonal change is triggered by the shortening of the days during the late summer.

is the length of the day rather than the length of the night that is measured.

The ability to detect the time of year enables living organisms to anticipate the onset of adverse conditions and make appropriate preparations for their survival. Swallows do not wait for the first cold weather in Northern Europe before they depart for a warmer climate, any more than trees wait for the first frost before forming their winter buds. Neither would survive for long if they did. Possession of a biological clock therefore enables many animals and plants to inhabit areas of the earth in which they would otherwise be unable to survive.

The easiest way to understand the biological clock and how it operates in plants is to first consider in some detail the characteristics of circadian rhythms, since these will reveal the essential clock-like features of the underlying oscillating system.

Measuring time

There are two types of mechanism with which time can be measured. The first involves the hour-glass or egg-timer principle in which a given quantity of sand runs from one chamber of a glass container to another in a known time. The mechanism then has to be reset to measure a further interval. The earliest water clocks were based on this principle, as were marked tallow candles which burnt down at a particular, known, rate. So far as is known, no plant or animal uses this principle to measure time. The second method involves an oscillation, like that of a pendulum, the escapement wheel of a small watch, or even, nowadays, the electrons of an atom like caesium. The oscillator has to be sustained by the continuous input of energy from a weight or spring and, of course, for accurate time-keeping the oscillations must be regular. This means that the period of oscillation, or the time between successive swings, must not vary. Achieving such regularity in the swing of the pendulum was a big problem with early clocks, especially with long-case or 'grandfather' clocks. The pendulum was usually a metre long, with a swing time of about one second, but the materials used in making the pendulum expanded and contracted as the temperature changed. On hot days the pendulum would be longer than on cold days, so the clock would be slow in summer and fast in winter! This problem was only overcome when the temperature-compensated pendulum

was invented. This type of pendulum was made of two different metals, with different coefficients of expansion, and was constructed in such a way that its overall length did not change with variations in temperature. The period of its swing was therefore constant, providing a sound basis for accurate time measurement. Detailed studies of circadian oscillations in plants and animals have shown that their periods also show a very high degree of temperature compensation, and so they possess the critical feature without which they could not function as biological clocks. None of the other rhythms in living organisms appear to have temperature-compensated periods.

Circadian rhythms

Circadian rhythms have been observed in a number of aspects of plant behaviour, such as stem growth, leaf movement and biochemical reactions, and these rhythms occur in both single-celled and multicellular organisms.

The beautiful single-celled alga *Acetabularia* has a circadian rhythm in its photosynthetic capacity. This means that when the cells are kept in a uniform environment, their ability to carry out photosynthesis changes in a rhythmic way over a 24-hour period. The rate of photosynthesis of a cell is not, therefore, solely dependent on the light intensity, temperature and CO_2 concentration, but is also regulated internally by the biological clock. A spectacular circadian rhythm of luminescence occurs in a single-celled, photosynthetic, marine organism called *Gonyaulax polyedra*. Many living organisms, such as fire-flies and the alga *Nostoc*, emit photons of light, but in *Gonyaulax* luminescence is obviously controlled by a biological clock since it will continue in a precisely rhythmic manner for weeks in laboratory cultures kept under uniform conditions.

Many fungi discharge their reproductive spores in a rhythmic way; some with explosive violence. The fungus *Pilobolus sphaerosporus*, for example, which grows on horse dung, produces a beautiful sporangiophore about five millimetres long. The black sporangium is formed on the end of the sporangiophore, which develops a large swelling just beneath the sporangium. This swollen vesicle must be highly pressurized, since the sporangium is eventually blown off with such force that it can travel for several metres. The sporangium is very sticky, so a strip of paper drawn over a population of sporangiophores at a constant

The single-celled marine organism *Gonyaulax polyedra* has a circadian rhythm of light emission from minute particles called scintillons in its cells (*right*). Large numbers of *Gonyaulax* cells (*far right*) give off a surprising amount of light – but they do so only at certain times of the day.

The fruiting bodies of the dung fungus *Pilobolus sphaerosporus* consist of transparent swollen sporangiophores with shiny black sporangia on their tips. A circadian rhythm governs the discharge of the sporangia. They are blown off with explosive force, but only at specific times of the day, and the rhythm remains unchanged even when the fungus is kept in controlled, constant conditions.

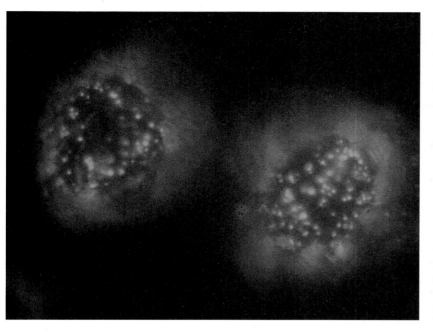

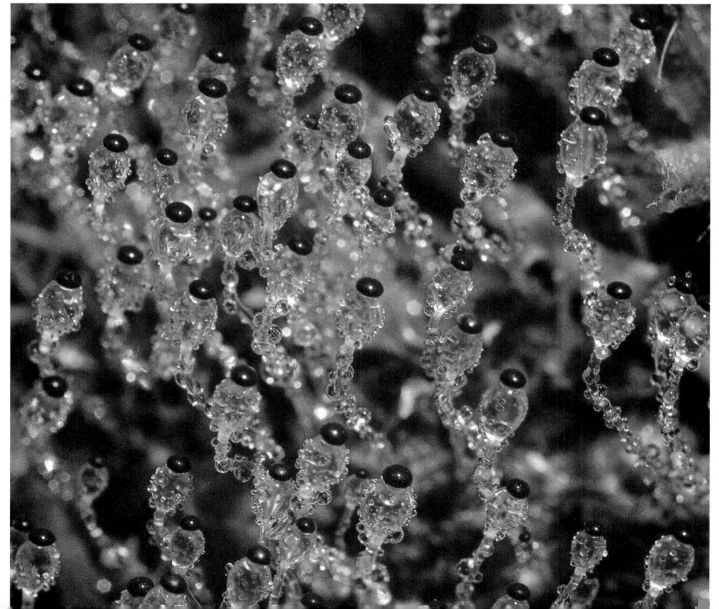

rate very soon reveals the rhythmic nature of the discharge.

Many different rhythms have been found in the flowering plants, a particularly well-studied one being the periodic movement of the leaves in plants like the runner bean (*Phaseolus coccineus*), *Albizia* and *Mimosa*. The leaf movement is brought about by changes in the hydraulic pressure in the cells of the pulvinus, located at the base of the leaf. When the cells in the lower half of the pulvinus are at maximum pressure the leaf blade is forced up into its more-or-less horizontal daytime position. Towards the end of the day, however, the pressure in the pulvinus cells decreases: the cells are no longer able to support the weight of the leaf, which then hangs down in the night-time position. The rhythm of the leaf movement therefore resolves itself at once into a question at the cellular level: what controls the circadian oscillation in the hydraulic pressure of the pulvinus cells? Is the biological clock located in these cells, or is it located elsewhere, with some means of sending signals to the pulvinus cells? In a plant called *Samanea*, rhythmic movement of the pulvinus will continue even if the leaf blade is cut off and the lower end of the leaf stalk immersed in a sugar solution – a discovery that showed that the oscillation in hydraulic pressure is generated in the pulvinus cells themselves.

Circadian rhythms also occur in the growth rate of shoots of cereals such as oats, in the opening and closing of the flowers of the succulent house plant *Kalanchoe blossfeldiana*, and in many metabolic processes such as the rate of carbon dioxide assimilation in the succulent plant *Bryophyllum fedtschenkoi*.

The leaves of the French bean (*Phaseolus vulgaris*) are held at different angles during the day (*far left*) and night (*left*). The leaf movements are caused by the expansion and contraction of cells in the lower part of the dark green pulvinus located between the top of the leaf stalk and the leaf blade.

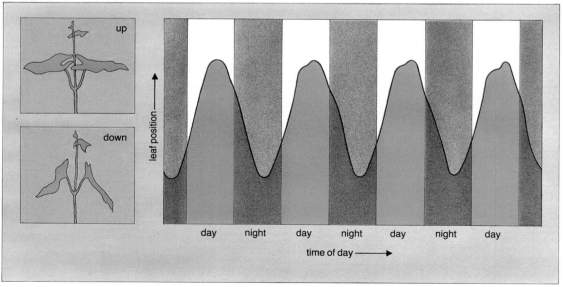

When plotted as a simple graph, the leaf angle of the French bean plant illustrates a typical circadian rhythm. The leaves begin to rise long before dawn, and start to droop again before dusk. This rhythm continues even when the plants are placed in permanent darkness and at constant temperature.

The biological clock mechanism

Underlying all these rhythms there must be a basic oscillating mechanism, and the first thing that we must do is establish its main characteristics. This is most easily achieved by asking the same questions that you would ask of a watchmaker if you were buying an expensive watch. These questions would be 1) How do you start the watch? 2) How do you adjust it? 3) How accurate is it? and 4) How often does it have to be wound up (or require a new battery)? These questions, translated into biological terms are 1) How do you initiate a circadian rhythm? 2) How can you reset its phase, that is, the time at which peaks or troughs in the cycle occur? 3) How well is the oscillation temperature-compensated? and 4) What is the energy requirement of the oscillator?

Circadian rhythms can be initiated in young, non-rhythmic seedlings by a single stimulus, which can be something as simple as just transferring the seedlings from light to darkness or vice versa, or exposing them for one or two hours to a different temperature. The importance of this finding is that it shows that the *period* of the oscillation must be an inherited characteristic, and that all the stimulus does is set the oscillator in motion. The oscillator can not acquire information about a periodicity from a single stimulus. A pendulum hanging at rest in a grandfather clock can be set in motion with a definite fixed period of oscillation by a single touch of a finger, and this is exactly analogous to the situation in a biological clock. It is quite impossible to teach biological clocks a new periodicity. A plant can be forced to oscillate faster by exposing it to, for example, 16-hour cycles of 8 hours light and 8 hours darkness, but no matter for how many days this treatment is continued, the plant will revert to its natural periodicity of about 24 hours as soon as it is placed in a constant environment. In the same way, a grandfather clock pendulum can be made to oscillate quickly by keeping hold of it and swinging it back and forth rapidly. It will never retain this forced periodicity however, and returns to its natural period immediately you let go.

Having started the oscillator, the next question is whether or not its timing can be adjusted. The answer is yes. When the oscillator is operating under constant environmental conditions its peaks and troughs will occur at particular times. Now, suppose we want to adjust the times of the peaks to occur at mid-night instead of midday. To achieve this, a stimulus has to be applied to the plant, and this stimulus can take the form of raising or lowering the temperature for an hour or so; exposing the plant to light for a similar time if it is in constant darkness, or to darkness if it is in constant light; depriving the plant of oxygen to bring respiration to a halt for a few hours; or treating the plant with certain chemicals. Each of these treatments will induce a phase shift in plant rhythms, the precise details of which will depend upon the plant being studied and the type of stimulus used. Each could be used, for example, to adjust the time at which the leaves attain their highest position in the cycle of leaf movement, and once the phase has been reset like this the leaf-position high spot will continue to occur at the new time.

To illustrate how important these studies are in attempting to understand the basic mechanisms involved, we shall consider the effect of a short, high-temperature treatment on the cir-

Leaves of the succulent plant *Bryophyllum fedtschenkoi* have a circadian rhythm in their output of carbon dioxide. The phase of the rhythm in continuous darkness can be shifted by a single 4-hour exposure to light given *between* peaks in the cycle (*upper graph*), but not by one given at the top of a peak (*lower graph*). On the graphs, the position of midnight is shown by zero.

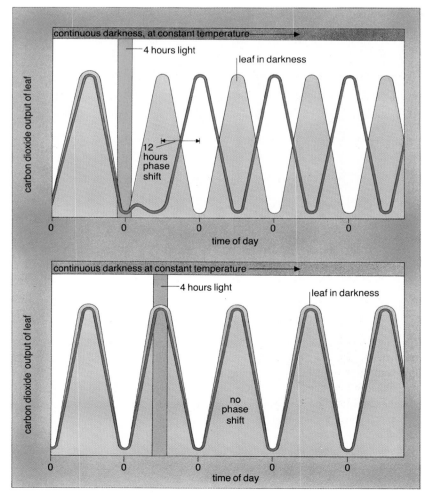

cadian rhythm of CO_2 uptake by leaves of the succulent plant *Bryophyllum fedtschenkoi* when it is kept in a stream of air at a constant 15°C and in continuous illumination. If the temperature is increased for three hours to 40°C, between the peaks of the rhythm, a large phase shift is induced: in fact, the rhythm in the treated leaves is completely out of phase with that in the untreated leaves. However, if the temperature is increased at a different point in the cycle, let us say, at the top of a peak of the rhythm, it is found that no phase shift is induced at all! So it is obviously not correct to say that increasing the temperature to 40°C for three hours shifts the phase of the rhythm. It does sometimes, but it depends on *where* in the cycle the increase in temperature occurs.

Studies such as this, in which the effect of the increased-temperature treatment has been determined over the whole cycle, have shown that at one point in the cycle, near the top of a peak in the rhythm, no phase shift is induced, whereas at all other parts of the cycle a phase shift is induced. The size of the phase shift changes with the position of the treatment and is clearly quite unrelated to the length of the treatment. What does determine the size of the phase shift is the exact moment when the high-temperature treatment ends, because the first peak of the new rhythm always occurs a specific length of time after this. This is a most important finding because it shows that during the exposure to high temperature, the oscillator does not merely stop, but is actually forced to, and held at, a characteristic and fixed point in the cycle. Thus, when the temperature is lowered to its original value again, oscillation *always* begins from this particular point in the cycle.

Very similar results are obtained with short exposures to low temperatures of 2°C. However, it is clear that at this temperature the basic oscillating mechanism is driven to, and held at, a fixed point in the cycle which is quite different, in fact 180° different, from that which results from a high-temperature treatment. In *Bryophyllum* leaves, therefore, low-temperature treatments have no effect on the phase when applied between the peaks, but are maximally effective at the tops of the peaks.

Similar results have been obtained with brief exposures to darkness and to light. When a rhythm is proceeding in continuous darkness, its phase can be reset by a three-hour exposure to light in a way that is virtually identical to

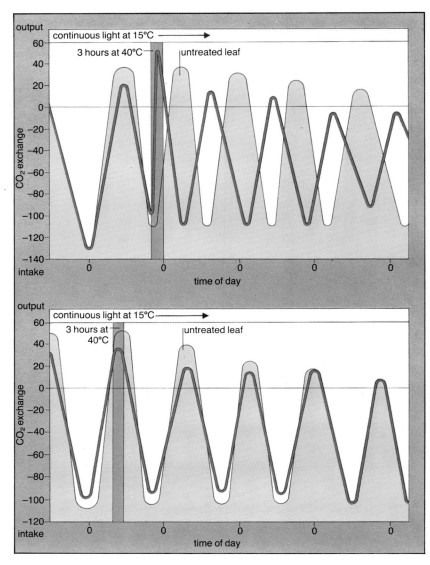

that achieved with a high-temperature treatment. In such a response to light, it is, of course, of interest to identify the photo-receptor involved. In higher plants there is no doubt that it is phytochrome, because only the red spectral band is active (Chapter 8). In rhythms in the fungi, on the other hand, only blue light is active, and the pigment is almost certainly a membrane-bound flavin similar to the one involved in phototropism (Chapter 7).

The third question we posed concerned the accuracy of the biological clock. If an oscillator is to be useful in measuring time with any degree of accuracy, its period of oscillation must not vary with changes in environmental conditions. In other words, it must show a high degree of temperature compensation, and this is true of all circadian rhythms so far investigated. The rhythms are not, however,

The phase of the circadian rhythm of CO_2 uptake by *Bryophyllum* leaves kept at 15°C in continuous light can be shifted by a 3-hour exposure to a temperature of 40°C administered *between* the peaks, but not by treatment administered at the top of a peak in the rhythm.

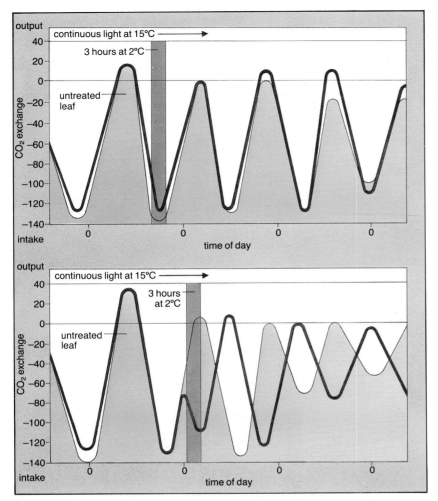

The phase of the rhythm in *Bryophyllum* leaves can also be reset by a brief exposure to a low temperature of 2°C, but in this case there is no effect when the treatment is given between the peaks, and maximum effect when it is applied at the top of a peak.

perfectly compensated because they do vary slightly with ambient temperature. The cycle of CO_2 output in *Bryophyllum* leaves kept in a stream of CO_2-free air and in darkness, for example, is 22.4 ± 0.4 hours at $26°C$ and 23.8 ± 0.3 hours at $16°C$. This difference is statistically significant, and the 'temperature quotient' achieved by dividing the period at $26°C$ by the period at $16°C$ is 0.94. This figure is very close to 1.0. If it had been 1.0, there would have been no change in the period at all and the rhythm would have been perfectly temperature-compensated. What makes the nearness of the value of 0.94 to 1.0 so interesting is that the oscillating system must consist of a series of biochemical reactions, each involving an enzyme and perhaps a co-factor. All chemical and biochemical reactions other than the photochemical reactions double, triple or even quadruple their rate when the temperature is raised by $10°C$. So the temperature quotient for these reactions will have a value of between 2.0 and 4.0! Somehow, the

reactions that generate circadian rhythms have incorporated a mechanism for combatting this effect. It almost seems that one normal dark reaction, whose rate increases with an increase in temperature, is balanced by another reaction whose rate *decreases* with temperature; there seems to be no other explanation for the high degree of compensation found in circadian rhythms. However, no one has yet identified any of the reactions involved in generating circadian rhythms or in achieving temperature-compensation of the period.

The final question was that relating to the energy requirements of the biological clock. All oscillators require energy: clocks have to be wound up to provide them with a store of energy in the spring or in the raised weight, or they must be provided with a supply of electrical energy in the form of a battery. There are two types of oscillator from the energy point of view; the pendulum, in which only part of the swinging weight's energy is lost in each cycle, and the relaxation oscillator, in which all the energy gained in one part of the cycle is lost in the other half. The repeated raising and lowering of a weight is an example of this kind of oscillator. The energy gained by the object during the raising part of the cycle is totally lost again when it returns to the start position. The biological clock or circadian oscillator certainly requires energy, and the energy is provided by the process of respiration (Chapter 12). If plants are deprived of oxygen, or if their respiration is hindered in some other way, then the circadian rhythms stop. If oxygen is withdrawn only for some hours, and then made available again, the rhythms will reappear in a new phase, indicating that the basic oscillator has, in fact, been stopped. Treatments involving depriving plants of oxygen for between three and six hours at different parts of the circadian cycle have shown that in some parts of the cycle the oscillation is stopped because a phase shift is induced, whereas in other parts of the cycle the phase is unaffected, indicating that oscillation has not been affected. This requirement for energy in one part of the cycle and not in the others is characteristic of the relaxation type of oscillator, and means in effect that the biological clock has to be wound up every day! A good deal more needs to be known about the energy requirements of circadian rhythms, however, before we can be certain that this analysis of the situation is correct.

Where is the biological clock – and what does it look like?

The biological clock exists at the cellular level, and its operation probably requires the cell to be intact and undamaged. There are many rhythms in single-celled organisms, the rhythms of luminescence in *Gonyaulax* and of photosynthesis in *Acetabularia* being excellent examples. Even in multicellular leaves like those of *Bryophyllum,* rhythms can be found in small pieces of tissue cut out of any part of the leaf, and even in cultured tissue from the leaves – a clear indication that each cell of the leaf possesses its own circadian oscillator.

An attractive place for the clock to be located within a cell would, of course, be the nucleus, since there it would be able to control enzyme synthesis and hence a very wide variety of chemical reactions and other cell processes.

Some extremely delicate experiments on the rhythm of photosynthesis in the umbrella-shaped single-celled alga *Acetabularia* were addressed to this question. Because the cell is very large, being one to two centimetres in length, with the nucleus located at the lower end of the stalk, it is possible to put a ligature around the stalk just above the nucleus and then cut off the portion containing the nucleus to provide an enucleate *Acetabularia* cell. In such cells the circadian rhythm of photosynthesis continues as though nothing has happened, so the nucleus appears not to be the site of the circadian oscillator. The oscillator must be in the cytoplasm, or associated with one or more of the cytoplasmic particles such as the chloroplasts or mitochondria. But a further experiment showed that although there was a functional circadian oscillator in the non-nuclear part of the cell, there was also one in the nucleus, and that the one in the nucleus could control the one in the cytoplasm! This was established by an experiment involving nuclear transfer. Two cultures of *Acetabularia* cells were kept under artificial day/night conditions in which the cycles were in opposite phases; while one was in the dark the other was illuminated. Nuclei were then removed from some of the cells in each culture and implanted into nuclear-free cells of the other culture: in other words, cells from the different light and dark cycles had their nuclei swapped over. Both cultures were then placed in constant environmental conditions and their circadian rhythms of photosynthesis were monitored to see if the phase remained the same in the

normal cells and in those that had received nuclei from cells maintained on reversed day-night cycles. In fact, the phase of the cells that had received a nuclear transplant was found to be completely the opposite of that in the untreated cells, and corresponded with the phase of the rhythm in the cells from which the nuclei had been originally taken. So a nucleus of *Acetabularia* does have a circadian oscillator or clock, and furthermore it can dominate or bring into harmony with its own phase all the other oscillators in the cytoplasm. It appears, therefore, that the nuclear clock is a master clock which controls the cell when it is present.

In the *Gonyaulax* luminescence rhythm there is evidence that the generation of the circadian oscillation may be associated with nuclear processes that copy the genetic information from the chromosome and use it in the synthesis of enzymes. The amount of the enzyme which causes the emission of light, luciferase, varies in a *Gonyaulax* cell throughout the circadian

These umbrella-shaped cells of *Acetabularia* have a circadian rhythm in their capacity to carry out photosynthesis. Each cell has a large nucleus located near the base of its stalk, and as this nucleus can easily be isolated and removed, the plant is very useful for demonstrating certain aspects of the 'biological clock'.

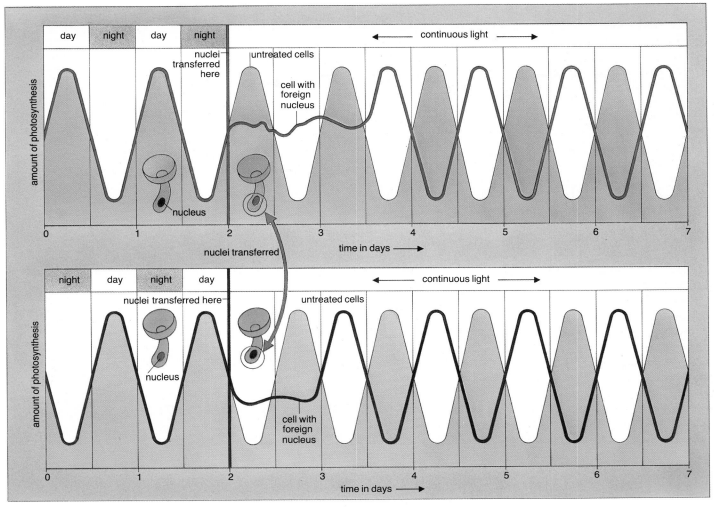

Figure labels (top panel):
- day | night | day | night | continuous light
- nuclei transferred here
- untreated cells
- cell with foreign nucleus
- nucleus
- amount of photosynthesis
- nuclei transferred
- time in days →
- 0 1 2 3 4 5 6 7

Figure labels (bottom panel):
- night | day | night | day | continuous light
- nuclei transferred here
- untreated cells
- cell with foreign nucleus
- nucleus
- amount of photosynthesis
- time in days →
- 0 1 2 3 4 5 6 7

When the nucleus of an *Acetabularia* cell is removed, both the cell and the nucleus retain their original rhythm when placed in continuous light. But if the nuclei of cells with opposite day/night rhythms are exchanged (green arrow), the rhythms of the recipient cells change and adopt the rhythms of their new nuclei – proving that the biological clock in the nucleus is the dominant one.

cycle of luminescence, as does the compound luciferin, also a protein, which it breaks down in the light-emitting reaction. Both these proteins therefore appear to be synthesized and broken down rhythmically, and this alone could give rise to the rhythm of luminescence.

In other higher plants it seems less likely that the circadian oscillation is generated in the nucleus, and more likely that it involves the movement of ions between different parts of the cell. Indeed, the changes in the hydraulic pressure in the cells of the pulvinus in *Samanea* and *Albizia* appear to be due to the loss and subsequent re-uptake of potassium (K^+) and chloride (Cl^-) ions. When these ions leak out of the cells through the plasmalemma on the lower side of the pulvinus, the hydraulic pressure in the cells declines and the weight of the leaf ensures that it begins to droop. Later, the cells on the lower side of the pulvinus take up K^+ and Cl^- ions again, their hydraulic pressure increases, and the leaf is returned to its

day-time position. No one has yet identified the mechanism underlying these ion movements, but periodic changes in the chemical structure of the plasmalemma so that it periodically becomes 'leaky' to K^+ and Cl^- ions may well be involved. But what controls so precisely the timing of these changes in membrane permeability is quite unknown.

The circadian oscillators in plants have not been identified in precise terms, but their characteristic properties are now well known. The nature of the oscillation, especially its temperature-compensated period, is fully compatible with it serving as the basic time-measuring mechanism or 'clock' in plants. In animals, the basic oscillation underlying circadian rhythms likewise remains unidentified, despite its practical and economic importance to man in the phenomenon of jet-lag, the loss of work efficiency after shift changes in factories, and the control of ovulation in agricultural animals such as chickens and sheep.

CHAPTER 15

Control of Flowering

In a great many plants, especially those of temperate regions, the onset of flowering, or the reproductive stage, is controlled photoperiodically; that is, by the length of the day or night. Such plants can be divided into a number of categories: those that flower in what are called 'short' days are called short-day plants; those that flower in 'long' days are called long-day plants, while those in which flowering is not controlled by day length are called day-neutral plants. In addition, a number of plants require more complex day-length conditions in order for flowering to be induced. Some require 'long' days followed by 'short' days, and are known as long-short-day plants, while others require the day-lengths to be in the reverse order, and are called short-long-day plants.

Among the best-studied short-day plants are the soya bean, the cocklebur, the succulent plant *Kalanchoe blossfeldiana* and the tobacco plant. Long-day plants include henbane and the culinary herb peppermint. Long-short-day plants include several species in the bryophytes, while the common clover is one of the most familiar of the short-long-day plants.

Photoperiodic control of flowering was discovered in studies of the late-flowering 'Biloxi' variety of the soya bean, which was found always to flower at the same time of year regardless of the date the seeds were planted. This clearly indicated that an environmental signal was involved, since at the time of flowering the plants could be anywhere between 50 and 125 days old. Other studies, on the 'Maryland Mammoth' variety of the tobacco plant, revealed that these plants would grow to an enormous size in the summer months when the days were long, but that they never flowered. However, when the plants were propagated in a greenhouse in the winter they flowered profusely, even though the plants were very small. Flowering evidently had nothing to do with the size of the plant, and it was soon established experimentally that in both these important American crop plants the sole factor controlling the onset of flowering was the number of hours per day for which the plants were exposed to light and to darkness.

Photoperiodic control of flowering is easy to understand if two important points are appreciated. First, it is not the length of the day that matters, but rather the length of the night. Unfortunately, the terminology relating flowering response to the length of the day was introduced before the importance of night length was established. A short-day plant is therefore really a long-night plant, and a long-day plant is really a short-night plant! Second, the actual length of the day or night at which the change-over to the flowering state takes place has no relevance to whether a plant is called a long-day or short-day plant. Flowering depends on what is called the *critical day-length* and this may be longer in a short-day plant than in a long-day plant. What determines whether a plant is called a long-day plant is that flower development is induced only when the length of the day *exceeds* the critical day-length, regardless of the latter's actual value. In the henbane, the critical day-length is about 11 hours, and the plant flowers only when the day-length is longer than 11 hours, never when it is shorter than 11 hours. Henbane, is there-

Chrysanthemum morifolium plants (*right*) will flower only when they are subjected to what are called 'short' days – that is, to days with less than a specific number of hours of light in each 24-hour period. The precise number of hours in this critical day-length varies considerably from one variety of the plant to another.

The succulent houseplant *Kalanchoe blossfeldiana* will flower only in short days (*right*). If kept in long-day conditions (*far right*) it will grow for years without ever producing flowers.

Hibiscus syriacus is a long-day plant. It will flower (*right*) in days longer than its critical day-length but never in short days (*far right*).

fore a long-day plant. In contrast, the cocklebur has a critical day-length of about 15.5 hours, and develops flowers only when the hours of daylight are *less* than this value. Cocklebur is, therefore, a short-day plant.

The realization that it was the night-length that was important in regulating flower development came from two types of experiment carried out in artificially illuminated controlled-environment rooms. In the first, the lengths of the day and night to which short-day cocklebur plants were exposed were varied independently so that the plants were no longer subjected to 24-hour cycles. Cycles of four hours of light and eight hours of darkness did not cause

flowering, even though four hours is much less than the critical day-length of 15.5 hours. Moreover, the plants in another room did flower when they were exposed to 16 hours of light and 32 hours of darkness, even though the light period was longer than the critical day-length. The length of the light period is thus irrelevant: the decisive factor is the length of the dark period. The first group of plants was exposed to eight hours of darkness, which was less than the critical night-length for cocklebur of 8.5 hours, whereas in the second group the imposed hours of darkness exceeded this value. So it is neither the day-length, nor the relative lengths of the day and night that

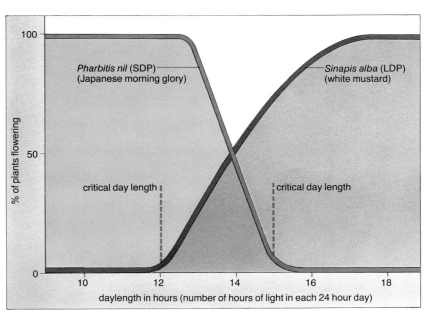

This diagram illustrates the response of a short-day plant, the Japanese morning glory (*Pharbitis nil*), and a long-day plant, the white mustard (*Sinapis alba*), to different day-lengths. Note that the LDP flowers only when the day-length increases beyond a critical value, while the SDP flowers only when it falls below a critical value.

control flowering in these plants; it is the *absolute* length of time for which they are exposed to darkness. And the dark period must be uninterrupted, because exposing plants to even a few minutes of quite dim light during the night can totally alter its effect. These so-called night interruption experiments have been of special interest for two reasons; not only have they confirmed the importance of night-length as a controlling factor in flowering, but by interrupting the hours of darkness with 15-minute exposures to light of different wavelengths (colours) it has been proved conclusively that the photo-receptor pigment involved in the photoperiodic responses of plants is phytochrome.

The measurement of night length by plants involves the measurement of the length of time that elapses between two light stimuli, dusk and

dawn, both of which activate the phytochrome pigment system. When darkness starts it seems that a circadian oscillator begins to operate, and it is the time in this cycle at which the second light stimulus (dawn) is received that measures the hours of darkness that have elapsed. This was revealed by experiments in which plants of *Chenopodium rubrum* were grown under continuous incandescent light and then placed in darkness for 72 hours before being returned to continuous light. *Chenopodium* requires only one night to be longer than its critical value for flowering to occur. In a series of experiments, different groups of plants were exposed to a few minutes of red light at different times during the 72-hour dark period, and the effect on the number of flowers induced was assessed. The results left no doubt that the flowering of the plants is regulated rhythmically according to the precise time at which the red light treatment was applied. In other words, there is a circadian rhythm in the flowering response of the plants to the red light treatment: at some times they flower and at other times they are prevented from doing so. This type of experiment clearly indicates that plants measure night-length by means of a circadian-oscillator-based timing mechanism, and by doing so they can detect the time of year and initiate production of their reproductive structures as well as control other important development processes. Now, flowers develop in the buds at the apices of the main and lateral stems, so we must try to find out precisely *where* the night-length measuring mechanism is located. Is it, for example, located in the bud in whose development it plays such a critical role?

Just where in the plant the night-length

Whether or not a plant will flower depends not only on the length of the night but also on any interruptions in the dark period. The colour bars (*right*) show how SDPs and LDPs react to various different treatments. From top to bottom the bars show: long days; short days; then short days with their night periods interrupted by a one-hour period of white light; by a one-hour period of red light; and finally by an hour of red light followed by an hour of far-red radiation. The last two bars show that the responses are controlled by phytochrome.

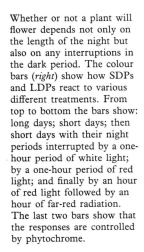

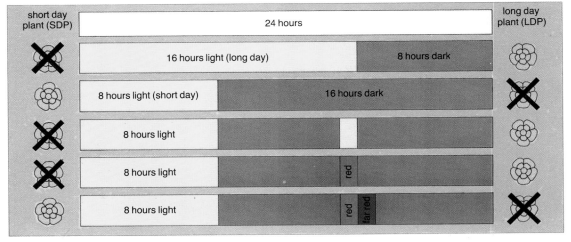

| short day plant (SDP) | 24 hours | long day plant (LDP) |

157

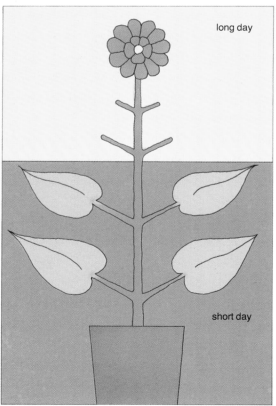

long day

short day

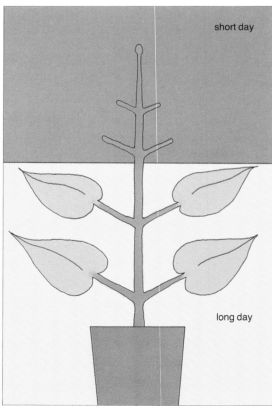

short day

long day

Chrysanthemum morifolium is a short-day plant in which the flower is really made up of a large number of very small individual flowers packed closely together on the flattened end of the flower stalk. The critical day-length is 11 hours in some varieties and as long as 16 hours in others.

To determine which part of the plant is responsible for measuring the length of the day, two *Chrysanthemum* plants have all the leaves removed from the upper part of the stem. They are then placed in divided chambers in which the upper and lower parts of the plants can be kept in different day-length conditions. The plant flowers only when the leafy part is subjected to short days – even if the apex, where the flower forms, is in long days. Measurement of day-length must therefore be taking place in the mature leaves on the lower stem.

measuring system is located has been revealed in experiments involving exposure of different parts of a plant to different day-lengths, wholly or partially removing foliage leaves from plants, and carefully grafting leaves from one plant onto another.

The first experiment was carried out on chrysanthemum, a plant that flowers only in short days. All the leaves were removed from the upper part of the shoots of a number of plants that had been grown in long days and were therefore in the vegetative state. These plants were then placed in special containers in which the top, leafless, part of the shoot was subjected to one day-length while the lower part, bearing the mature foliage leaves, was subjected to another. Subjecting the leafless apex of the shoot to long days and the lower part to short days induced flowers to develop in the apical bud, whereas the reverse treatment caused no flowers to form. This finding is important for two reasons. First, it shows that day-length measurement does not occur in the apical bud, where the flowers actually develop, but rather in the mature leaves. Second, it shows that there must be a transmission of some flower-inducing stimulus or signal from the mature leaves up through the stem to

the apical bud where flower development is initiated. The question then arises of whether a specific flower-inducing hormone is produced in the leaves and for which the target cells are in the active meristem in the apical bud. We shall return to this question later, but first we need to look more closely at the day-length detection mechanism in the leaves, to see whether it will tell us more about the nature of the signal transmitted from the leaves to the apical bud.

In the cocklebur plant, which flowers only in short days, it is possible to expose only one of the many leaves on a plant to short days, while keeping the remainder in long days. When this is done, the plant flowers. Furthermore, a single leaf can be detached from a cocklebur plant that has been exposed to a few short days and grafted on to one that has been, and still is, kept entirely in long days. Since this plant then flowers, it is clear that the single leaf carries its flower-inducing message to, and releases it into, the host plant.

A question of some importance in the photo-periodic control of flowering is whether the leaves of a plant in non-flower-inducing day-lengths produce an inhibitor of flowering, which then ceases to be produced in inductive day-lengths, or whether nothing is produced in non-inductive day-lengths while a flower-promoting substance is produced in inductive day-lengths.

In short-day plants, several of the experiments already mentioned point strongly to the correctness of the second hypothesis, namely that, a flower-*promoting* signal is produced by leaves in inductive short-days. When all the leaves are removed from a short-day plant kept in long-day conditions it does not burst into flower, which might be expected if the leaves exposed to long-days were producing an inhibitor. Secondly, exposing a single leaf to short days, or grafting on a single leaf from a plant kept in short days, induces flower formation in a short-day plant even when all the other leaves are being exposed to long days. Decreasing the supply of a supposed inhibitor by the amount produced by a single leaf is unlikely to induce flowering: what is much more likely – and consistent with experimental findings – is that the one leaf exposed to short days, or grafted on to the plant, is producing a *promoter* of flowering while the rest of the leaves on the plant, which are exposed to long days, produce nothing which influences flowering in any way.

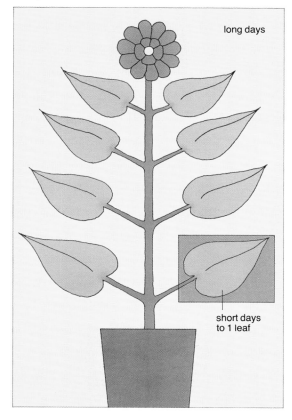

Even the slightest exposure to the correct day-length conditions can trigger the flowering response. In this experiment, exposing a single leaf of the short-day plant *Xanthium strumarium* to short days has produced a flower, even though the rest of the plant was kept in long-day conditions all the time.

long days

short days to 1 leaf

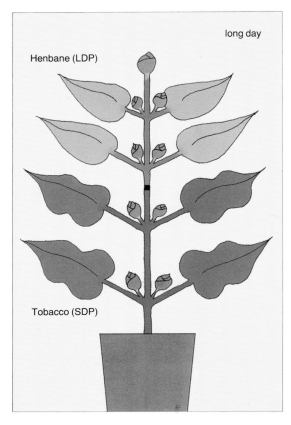

long day

Henbane (LDP)

Tobacco (SDP)

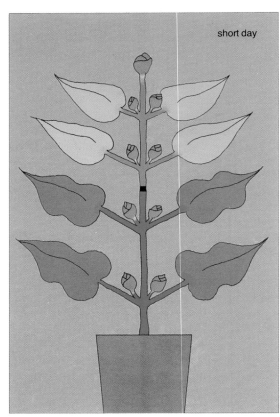

short day

If a shoot of henbane, a long-day plant, is grafted onto a shoot of tobacco, a short-day plant, flowers will develop on both shoots in both long and short days, since the composite plant is equipped to measure, and respond to, both conditions.

The single, induced leaf could have the effect of releasing into the plant a flower-initiating chemical, thus introducing a factor that was not there before – a situation that is much more likely to produce the precise control over flower development that obviously exists. To measure a flower-inducing day-length the leaf does not even have to be attached to the plant! If the short-day plant *Perilla* is kept in long days so that it does not flower, a single leaf may be detached, exposed to short days, and then grafted back on to the mother plant which promptly develops flowers.

The possibility of their being a universal hormone that induces flower development in both short-day and long-day plants is indicated by experiments in which plants with different day-length requirements have been grafted together. For example, the short-day plant cocklebur will flower in long days when it has been grafted on to the long-day plant *Rudbeckia bicolour*. The flower-initiating stimulus generated in the latter plant is obviously transferred to the cocklebur through the graft union. When the short-day variety of tobacco, 'Maryland Mammoth', and the long-day plant, henbane, are successfully grafted together, *both* plants flower in short days and in long days. The

flower-inducing signal produced by either plant can induce flowering in the other – a clear indication that the chemical signal must be identical in both cases, or at least very similar indeed.

An enormous number of fascinating and ingenious experiments have been carried out to try to unravel the problem of how flower initiation is controlled in photoperiodically responsive plants. Each species appears to be slightly different from the next, and what have been described here are just the more general findings. Trying to fit all the known facts into one unified mechanism is probably unwise – and incorrect. We have only considered, briefly, the 'simple' cases of short-day plants and long-day plants: remember that some plants have a requirement for both these day-lengths to occur in succession, and in the correct order, and that little or nothing is known about the mechanisms involved in these plants at present!

The mystery flowering stimulus

There can be no doubt that the signal emitted from a leaf in flower-inducing day-lengths is a chemical of some sort. In the early days it was anticipated that this was a specific substance whose sole function was to induce flower devel-

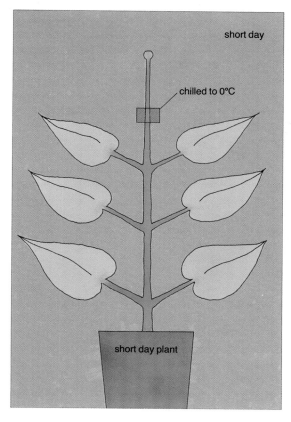

If part of the stem of a short-day plant is chilled, between the uppermost leaf and apical bud, the bud will fail to flower even if kept in short-day conditions. With its phloem system out of action, the plant cannot transport the flowering stimulus from the leaves to the apical bud.

short day

chilled to 0°C

short day plant

opment at the apical buds of the main and lateral shoots. This supposed flower-inducing hormone was even given a name – 'florigen' – but there is now considerable doubt about the existence of any such substance.

Most of the hormone-like chemicals in plants differ from the hormones found in animals in that they do not have a single, characteristic function in controlling a specific aspect of growth, development or metabolic activity. In plants, while it is true that each of the hormone-like groups of substances does have a characteristic activity possessed by none of the others, they each have other activities as well, either alone, or in combination with other molecules. This problem immediately opens up the possibility that florigen may not actually exist, and that the chemical signal emitted by an induced leaf may simply be a change in the normal pattern or balance of some of the other known groups of hormone-like molecules, especially as some of these can readily induce flowering in photoperiodically responsive plants.

The real argument against the existence of florigen is the fact that despite years of work by many plant biologists and chemists in laboratories throughout the world, no one has ever been able to extract an active substance from

an induced plant. To establish conclusively that florigen exists, it would be necessary to demonstrate that the chemical is present in induced plants and absent in non-induced plants, and also that when extracted from an induced plant, it will initiate flower development in a non-induced plant kept in non-inducing day-lengths. Neither of these requirements has been achieved, and in their absence it is impossible to begin the task of identifying the precise chemical structure of the compound. The conclusion to be drawn at present is that 'florigen' is simply a convenient name for the flower-inducing signal released by a photoperiodically induced leaf. There is no evidence that it is a single, hormone-like chemical substance; it may just be a particular 'mix' of several of the known growth-regulating chemicals produced by the foliage leaf following exposure to flower-inducing day-lengths.

Whatever the chemical nature of the signal, it travels quite a long way from the foliage leaf to the stem apex, and there is convincing evidence that its transportation depends on living cells, probably those of the phloem. If a part of the stem is chilled, to reduce the availability of respiratory energy, transmission of the stimulus is stopped and the apical bud does not develop flowers. Killing the living cells in a ring around the stem with a steam girdle has a similar effect. The stimulus can not, therefore, be carried up the stem in the water stream in the xylem, because this is dead tissue and its function would be unaffected by these treatments. The rate at which the stimulus passes up the stem in plants such as the Japanese morning glory is the same as that observed for the products of photosynthesis, about 30 cms per hour, and this suggests that the flower-inducing stimulus is transported in the living transport tissue of the phloem. An active transport system must be involved because the movement of the stimulus is much too quick to be accounted for by physical diffusion.

Vernalization

In a number of plants, usually long-day plants, there is yet another mechanism that regulates the onset of flower development. This mechanism is found, for example, in winter cereals and in biennial plants, and it involves a requirement for exposure to a low temperature of between 2°C and 5°C for an extended period of time, usually several weeks. It is this require-

All our root crops need to go through a period of cold, or vernalization, before they are able to flower. The red beet-root builds up food reserves in a swollen root during its first season of growth (*above*), but produces no flowers. During the following spring and summer, after exposure to the winter cold, these food reserves are used to produce the plant's huge flowering stem (*right*).

ment – known as vernalization – that distinguishes the winter cereals from the spring cereals, both of which are, in fact, annual plants. Spring barley or wheat is planted in the spring and flowers some weeks later. If winter varieties are planted in the spring they do not grow quickly enough to flower before autumn and so no crop is produced. If they are planted in October or November the seeds germinate and grow slowly to the small seedling stage, in which condition they remain during the cold winter weather. In the following spring, however, they grow and develop flowers very quickly indeed. In the 'Petkus' variety of rye, which has been studied in detail, it is clear that under ideal conditions the winter variety will eventually flower even if it has not been exposed to a cold winter period. However, the flowers develop only after the seedling has produced about 22 leaves, which can take between 14 and 18 weeks. Seedlings that have been exposed to 2°C for several weeks behave just like the spring variety but flower after developing only about six or seven leaves, which takes about seven weeks. Some varieties of winter wheat have what is known as an absolute requirement for a cold period: without it they remain vegetative and never develop flowers.

Despite its critical role in promoting flower formation, the effects of vernalization are not immediately apparent on the seedling itself. There are, for example, no recognizable flowers on the shoot apex at the end of the treatment. However, the chilling has performed its task of inducing the onset of the flowering process: the flowers appear later when the seedling is once more in its normal temperature of 15°C to 20°C.

The part of the plant where the cold treatment is detected appears to be the apical buds, the very place where the flowers eventually develop. This was established by experiments in which the buds themselves were chilled while the rest of the plant was kept warm; chilling other parts while keeping the buds warm did not work! Vernalization therefore differs from photoperiodism in that detection of the stimulus appears to occur at or very near the site of the response. No transmission of a signal from one part of the plant to another is necessary. Apical buds have even been cut off a plant and placed for several weeks in a refrigerator. When replaced on the stem apices of plants kept throughout that time at room temperature, the treated buds eventually developed flowers! For the most part vernalization is detected in areas of the plant where active cell division is taking place – in the apical meristems of the main and lateral shoots.

No one has yet established just what happens during the chilling period to bring about the onset of flowering. A number of new chemical compounds do appear after the seedlings have been chilled, and a changed pattern of proteins has been found in both rye and wheat. Some of these proteins may be critical for flower development because if seedlings are treated with inhibitors of protein synthesis the new proteins do not appear after a period of chilling and flowering is not induced. It is possible, of course, that a specific, flower-inducing, chemical compound or hormone could be produced after the chilling period, formed perhaps as the result of the 'de-repression' of a specific gene on the chromosome. If this gene carries the message to make an enzyme that can promote the synthesis of a special flower-inducing hormone, then if it is depressed, or switched off, no flowering would be possible. The cold treatment might thus lead to de-repression – the removal of the block which prevents the information on this gene being read. However, evidence for the existence of flowering hormones is very thin indeed, and so far, attempts to isolate and identify an active chemical stimulus in vernalized plants have met with the same marked lack of success as those aimed at identifying an active 'hormone' involved in the photoperiodic control of flowering.

It is all too clear, then, that very little is known about the biochemical mechanisms underlying the phenomenon of vernalization. Many of our difficulties stem from the fact that we have no knowledge of the really critical chemical changes that precede the natural development of flowers in plants with either photoperiodic or vernalization requirements, and until it can be established whether or not a specific flower-inducing hormone is involved there appears to be little prospect of progress. Such an understanding of these phenomena is of some importance since a large part of the world's temperate zone production of sugar, and of other root crops, depends on the fact that such biennial plants do not flower but rather store their food reserves during the first season of growth in order to provide energy for the second year's growth when the plant produces its flowers.

At the end of their first growing season, when they are harvested for human use, the turnip (*above*) and sugar beet (*left*) have built up large stores of carbohydrates in their swollen root structures. The sugar beet is particularly rich in sucrose.

CHAPTER 16

Plant Hormones

Many aspects of an animal's behaviour, growth and metabolism are regulated by special chemical molecules called hormones, which are essentially chemical messengers. They are defined as chemical substances formed in one part of the body and transported to other parts where they perform specific functions in regulating essential processes. They can start, stop, promote or suppress particular processes, and they are active when present in extremely small quantities. Their chemical structure varies considerably. Some, like insulin, which regulates the sugar level in the blood, are polypeptides – chains of amino acids joined end to end. Others, like the prostaglandins, are derived from quite different chemical structures – the fatty acids – while a further group, the steroids, are complicated organic molecules amongst which are the human sex hormones. The different chemical structures clearly regulate different processes in the animal body, and it is obvious that the target cells, where the hormone exerts its effect, must be able to recognize the arrival and the amount of a specific hormone. This implies that there are sophisticated sites in the target cells on to which the hormone must latch before it can exert its influence, just as a particular lock can be operated only by one particular shape of key.

In this chapter we shall consider whether plants have regulating hormones like animals, and, if so, what we know about their chemical structure and mode of action.

There is no doubt at all that plants do regulate their growth, development and behaviour by means of chemical substances which in most cases move through the plant from one part to another. In plants, the sites of synthesis of some of these chemical regulators are not always as clearly localized as they are in animals; neither are the target tissues. Furthermore, plant hormones do not appear to have individual, specific, activities; they often regulate many different and seemingly unrelated processes at the cell or tissue level. On the other hand, they are active in such extremely small amounts that it is possible to obtain a response by treating a plant with a solution containing one of these substances at a concentration of between one-millionth and one-ten-millionth molar – between 0.2 and 0.02 parts in a million!. So are we justified in calling these chemical regulators hormones? Despite the difficulties in making a direct comparison with animal hormones, it probably does no harm to call them plant growth hormones since they do function in plants in much the same way as do hormones in animals.

Although each group of plant hormones regulates a number of quite different activities, each, in fact, has at least one characteristic activity possessed by none of the others. This is perhaps not surprising because each group has a quite different chemical structure which evidently can be recognized in the target tissue. However, plant hormones may also interact with one another to control developmental processes, so that in some cases it may be the relative balance of two or even three hormones that is the critical factor.

Plant hormones are divided into four groups, with an additional single chemical substance – the gas ethylene. The four groups are the

A unique feature of plant cells is that each one contains the genetic code for the whole plant. Leaf cells are therefore able to give rise to roots and stems.

The *Streptocarpus* leaf (*opposite, top left*) has sprouted roots, and also a young stem from which new leaves are growing. The usual way of propagating *Begonia* plants (*top right*) is by detaching leaves and placing them on wet sand. Young plantlets soon grow around the leaf base.

Other plants, such as *Bryophyllum daigremontianum*, produce young plantlets spontaneously on the margins of their leaves (*below*). Eventually these drop to the ground and begin an independent life.

auxins, the gibberellins, the cytokinins and the inhibitors. The first three groups all have a characteristic chemical structure whereas the fourth group comprises a number of quite different chemical substances.

Auxins

When Charles Darwin carried out his famous experiments in phototropism in 1880 he found that if only the very tip of a grass coleoptile was illuminated from one side by a candle, curvature of the organ occurred much lower down, in a part that had not been exposed to light at all. Darwin wrote, 'we must therefore conclude that when seedlings are freely exposed to a lateral light some influence is transmitted from the upper to the lower part causing the latter to bend.' In referring to the movement of a growth-regulating 'influence', Darwin was the first to recognize the existence of a hormone-type control mechanism in the growth of plants. Many years were to pass before it was established just what this influence was, but in the meantime it was called auxin.

In the early part of the present century much interest was centred on the growth-controlling 'influence' operating in coleoptiles, and most of the experimental work was carried out on the coleoptiles of oat seedlings. It was quickly established that a growth-*promoting* substance was involved, because if the coleoptile tip was cut off and replaced asymmetrically, the coleoptile subsequently bent away from the side covered by the tip. It was later found that when a detached tip was placed for an hour or so on a small block of agar, the growth-promoting influence would diffuse out of the tip into the agar block. When the block alone was then applied to one half of the end of a freshly detipped coleoptile it induced curvature.

But it would be naive in the extreme to assume that only one growth-regulating compound was being released from the coleoptile tip. The promoting effect of the diffusate in the agar block could indeed be due to a single hormone – but it could also be due to two or three or more compounds, promoters *and* inhibitors, the final growth-promoting effect being determined by the relative amounts of the various substances. For this reason it is correct to speak only of the *net* growth-regulating influence of the diffusate. What is quite clear, however, is that there must be at least one-growth promoting substance present and we now have to establish its chemical identity.

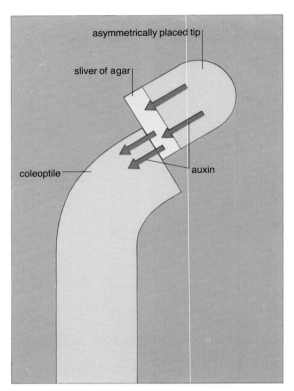

asymmetrically placed tip

sliver of agar

coleoptile

auxin

If the tip of a corn coleoptile is cut off (*left*) and stuck back on again asymmetrically using a thin sliver of agar to hold it in place, the coleoptile will bend as shown. This indicates that a growth-promoting substance, auxin, is produced in the tip, passes into the shoot, and causes the cells on that side to grow.

The diagram opposite shows the molecular structure of auxin, the chemical that controls plant growth. The chemical name for auxin is indole-3-acetic acid (IAA).

The invention, in 1949, of chromatography, an important method for separating chemical substances from one another, allowed extracts from plant tissue to be analysed with a precision that had not been possible before. Paper chromatography quickly revealed that there were several growth-regulating compounds present in plants, some promoters and some inhibitors, and that they were present only in very minute amounts.

The major growth-promoting compound in these studies always seemed to behave in much the same way as a known compound called indole-3-acetic acid (IAA) and it was assumed that auxin was, in fact, IAA. However, several other compounds were also known to behave in a similar way and so there was no proof that auxin and IAA were one and the same compound. (It is sometimes forgotten that chromatography is a separation procedure, not an identification procedure.) In the early 1970s, however, unequivocal proof was finally obtained that auxin was indeed IAA.

It had been known for many years that IAA could promote the elongation of cells. Segments of coleoptiles placed in very dilute solutions of IAA grew at a rate related to the concentration of IAA in the solution. Furthermore, IAA is almost unique in that it is transported in plant shoots and in roots in a polar manner, that is,

The promotion of cell growth in sections of oat coleoptile after 18 hours in a very dilute auxin solution (*lower picture*) is very obvious when compared with sections kept in water for the same length of time (*top*).

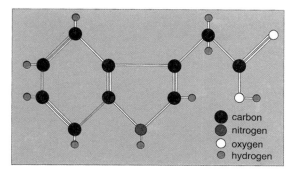

carbon
nitrogen
oxygen
hydrogen

in one direction only. In a coleoptile it always travels from the tip to the base, even if the tissue is upside down! In a root, IAA is transported only from the base to the tip. This polar transportation depends on metabolic energy since it comes to a halt if respiration is stopped by depriving the plant of oxygen. So IAA fulfils the role of a plant growth hormone in that it is produced at the tip of a coleoptile and transported to the lower regions where it controls the rate at which the growing cells actually elongate. It fulfils entirely the requirements of Darwin's 'influence'. IAA is the only compound occurring naturally in plants that seems to have the capacity, at extremely low concentrations, to regulate the rate of elongation of cells in the growing zone of coleoptiles. This, then, is the essential characteristic activity of auxin that is possessed by none of the other hormone groups.

The mechanism of cell growth

Detailed studies, spread over many years, have revealed that there are sites on the membranes in plant cells that bind IAA specifically and exclusively. It is the special features of the binding site that evidently recognize the IAA molecule. No other molecule will bind in its place. We do not yet know for certain whether all or only a few of these sites are involved in growth regulation, because it appears that there are several types of binding site which differ in their affinity for the IAA molecule. Neither is it clear whether the important sites are on the plasmalemma or on the other membranes in the cell. Nevertheless, having bound itself on to the membrane site, we must now establish how the IAA-site complex actually increases the growth-rate of the cell.

As we saw in Chapter 2, a plant cell wall is a very strong structure, rather like the case of a football. It is this outer case that constrains the large pressure built up inside the cell by osmosis. The cell wall is made up of chains of

Roots develop at the basal end of shoots, even when they are held upside down. The mass of roots at the basal ends of these two inverted bamboo shoots is thought to be due to the fact that auxin is transported in one direction only – that is, from the tip towards the base.

A section of dandelion root (*Taraxacum* sp.) will develop a shoot at the apical end (never the basal end) if it is kept in a humid environment for a few weeks. In roots, the auxin always travels from the base towards the tip.

carbohydrate polymers, the principal one being cellulose, which consists of long chains of glucose molecules. Other carbohydrate polymers are made of different sugars; sometimes only one sugar is involved, sometimes there are several. Cell wall polymers made of sugars such as xylose, mannose and galactose are called hemicelluloses, and they are usually present in smaller quantities than cellulose. In some algal cells the cell wall consists entirely of a hemicellulose such as xylan (a xylose sugar polymer) or mannan (a mannose sugar polymer). In higher plants, however, the main component of the wall is cellulose, with hemicelluloses making up most of the rest. There is, however, a small amount of protein in the wall, along with another important carbohydrate polymer made up of units of a sugar-acid called galacturonic acid. The polymer of this molecule is pectin, and its importance in plants is that it acts as the glue that sticks the cells together. Pectin is soluble in boiling water, which explains why there is such an awful mushy paste in the bottom of the pan if potatoes are boiled for too long. The cell's separate from each other and form a sludge because the pectin 'glue' which was holding them together has been dissolved away.

The various carbohydrate polymers are intertwined in the cell wall and numerous chemical bonds project outwards from the sides of these polymers. These bonds link the polymers together laterally and evidently prevent them from slipping past one another as the wall is stretched by the pressure generated within the cell. The cell wall has tremendous strength, and it is clear that if the cell is to grow, the wall must weaken in a controlled way. So what IAA does when it arrives and locks on to the receptor site in the cell membrane is to set in train a series of processes that ultimately lower the tensile strength of the cell wall. This appears to be achieved by breaking some of the lateral chemical bonds linking the carbohydrate polymers together, thus allowing them to slide past one another or to separate from one another.

The key to the process lies in the fact that making the cell wall acid reduces its resistance to stretching. What happens is that as the IAA molecules arrive in the cells they bind to the receptor sites and immediately initiate the operation of a pumping system that pumps hydrogen ions (H^+) from the cytoplasm, through the plasmalemma and in to the cell wall. Increasing the concentration of hydrogen ions in the wall increases its acidity, because acidity is actually a measure of the concentration of free hydrogen ions at a particular place. The higher the concentration of the IAA solution supplied to the plant, the higher will be the number of ion pumps brought into operation. The more acid the wall becomes, the less resistant it is to the stretching forces generated by the osmotic system of the cell. Precisely how increased acidity reduces the wall's tensile strength is not yet fully understood. One possibility is that the lateral bonds binding the carbohydrate polymers simply give way in an acid environment. Another, and more likely, explanation is that the increased acidity activates special enzymes in the cell wall, and that it is these enzymes that break the bonds and bring about the controlled reduction in the tensile strength of the wall.

The Instron stress/strain analyser above is being used to apply a known stretching force to a section of oat coleoptile. The instrument makes it possible to measure the resistance of the cell walls to stretching, and so investigate the processes of cell growth.

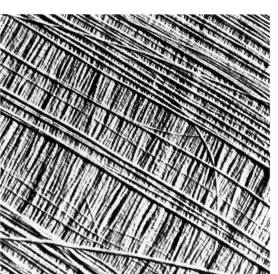

Far left: An electron microscope photograph reveals the microfibrils in the cell wall of a wheat coleoptile. The circular area in the centre is a pit forming in the cell wall.

Left: The cellulose microfibrils in two layers of the cell wall of the green alga *Chaetomorpha melagonium* are laid down almost at right-angles. Note that one of the microfibrils crosses over from one layer to the other.

Apical dominance

In addition to its unique property of promoting coleoptile cell growth, auxin has a number of other effects on the growth and development of plants. In the phenomenon of apical dominance the presence of an actively growing bud at the tip of a main stem suppresses the growth and development of lateral buds lower down the stem. It appears that the IAA being produced by the apical bud is responsible for, or at least involved in, the lateral bud suppression.

If the apical bud is removed from a runner bean plant, the two lateral buds begin to grow within a few hours. If, on the other hand, IAA is applied to the cut stem apex immediately after the apical bud is removed, the laterals do not grow. Inhibition of the lateral buds appears to require a continuous flow of IAA from the apex of the stem, because if the apical bud is

The paired pictures here illustrate the hormonal control of lateral bud growth. Those on the left provide overall views of the plant: those on the right are close-up views of specific details.

In the intact plant (*top*), lateral buds in the leaf angles do not grow. If the apical bud and main shoot are removed (*centre*) the laterals start to grow within hours. But if a capsule containing lanolin and IAA is placed on the cut end as soon as the apical bud is removed (*bottom*) the growth of the laterals is prevented.

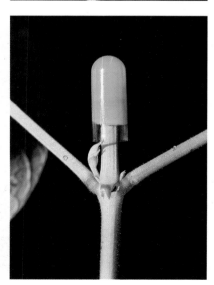

169

removed and the application of IAA delayed for several hours, the IAA no longer suppresses growth of the lateral buds.

Numerous ideas have been advanced to explain apical dominance, but the truth of the matter is that we are not sure, at present, of the nature of the signal transmitted from the apical bud to inhibit the laterals, or how it operates when it reaches its target cells. There is considerable evidence that IAA is involved and possibly other plant growth hormones as well, but whether IAA operates alone or in combination with one or more of the others has not been established. Despite the importance of apical dominance in horticulture and agriculture there is still a great deal to be learned about what happens when a gardener prunes his roses and other shrubs in the autumn and spring to encourage lateral bud growth and a sturdier plant with more growing shoots.

Rooting of cuttings

IAA is a strong promoter of root initiation, that is, it encourages roots to develop – even on stems and leaves. This is an extremely important matter because it forms the basis for all genetic engineering in plants (Chapter 18). IAA can be shown to promote root development by taking stem cuttings and placing the cut ends in a dilute solution of IAA while other cuttings are placed in water for comparison.

Rather few roots will appear in the water-treated cuttings whereas there is prolific rooting in the ones treated with IAA. It is important to recognize what is happening here – a stem cell or group of stem cells is being induced to make a root! The implication of this phenomenon is that the stem cells contain all the basic information on just how to go through the sequences of cell division, growth and differentiation to make a root, even though this information has never been required in the stem cells. The fact that plant cells clearly retain all the information required to make the rest of the plant is of immense importance.

Leaf and fruit drop

Leaves at the end of their active lives, and mature fruits when fully ripe, are shed from plants in a process known as abscission. These organs do not die on the plant and fall off; neither are they torn off. They are, in fact, cut off in a highly controlled process involving the breakdown of cells in a special thin layer across the leaf stalk or fruit stalk. This layer of cells is called an abscission zone, and the process has been best studied in the leaf stalks (petioles) of the popular house plant called *Coleus*.

A leaf approaching the end of its life seems to activate the cells in the abscission zone to produce enzymes that led to the breakdown of their cell walls, so that the structural integrity

The one-way movement of auxin in plant organs is thought to be the reason for the strict polarity in organ initiation. In the demonstration shown below left, roots have formed only on the upper edge of a hole cut in a *Sanseveria* leaf, because this is the basal edge of the tissues above the hole.

Leaves do not simply die and fall off in winter. They are cut off, very precisely, in a process called abscission. The middle picture below shows leaves that are dead but still firmly attached to a hawthorn branch cut from a tree in August and left on the ground till November. It was then photographed alongside a branch (*below*) that had remained on the tree throughout, and from which the leaves had been shed in the normal way.

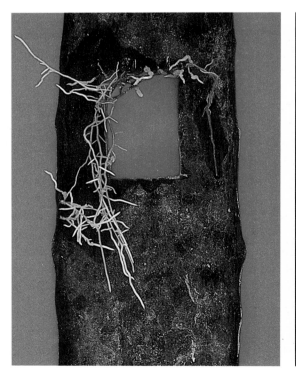

of the tissue totally disappears. The leaf is then left attached to the plant only by the few strands of xylem cells in the vascular bundle. These strands are quite brittle and easily break when the leaf is touched or moved by the wind or rain. Before the separation process is completed, however, the tissue of the mother plant next to the abscission layer produces a very localized cork cambium. When the leaf has fallen off, this impervious cork layer covers the living tissue that would otherwise be exposed, and prevents infection by fungi and bacteria. (A very similar process occurs when a tree suffers physical damage. The wound is quickly sealed to prevent infection.)

But what actually activates the abscission layer? A number of factors may be involved, but one appears to be IAA. If a leaf blade is removed, abscission of the remaining petiole is quickly induced, but this activity is very considerably delayed if IAA is placed on the cut end of the petiole. While IAA may be involved in controlling dissolution of the abscission layer cells, we do not yet know precisely what controls the onset of cellulase activity, the enzyme responsible for the breakdown of the cell walls. A number of other hormones such as the cytokinins and abscisic acid also appear to be involved, and the process may be controlled by the balance between a number of the plant growth hormones.

When leaves are shed from a deciduous tree, each one leaves a scar (*left* and in close up *below left*). The scar is sealed by a layer of impervious cork cells which prevent infection entering the wound. The ring of broken vascular bundles that ran out into the leaf stalk shows very clearly around the edge of the leaf scar. Immediately above is a lateral bud.

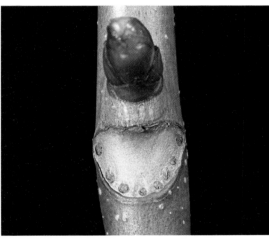

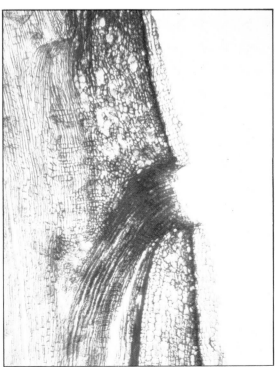

The microscope reveals the inside story of how abscission occurs. Before the leaf is shed (*far left*) a special layer of thin-walled (light coloured) cells across the basal end of the leaf stalk becomes active and the cells 'self destruct' by a process of digestion. This effectively detaches the leaf, except for the xylem tissue which snaps when the wind moves the leaf. Beneath this abscission layer is a layer of actively dividing cells which produce the protective covering of cork that seals the scar against infection. This can be seen (*left*) on a photograph taken after leaf fall.

171

The gibberellins

The gibberellins were discovered from studying the effects of a pathogenic fungus which infects rice plants. The fungus causes the rice plant to grow much taller than its uninfected neighbours; so tall, in fact, that it is unable to support itself, falls over, rots and hence produces no crop. This disease of rice plants is called the 'bakanae' disease, meaning 'foolish plant', and the fungus that causes it is called *Gibberella fugikuroi*. In 1926, it was discovered that this fungus caused the over-growth of rice plants because it produced a complicated molecule called a gibberellin. Later, it was found that if the fungus was grown in pure culture in a nutrient medium, the gibberellin was produced in quantities large enough for it to be chemically identified. More than sixty of these gibberellins are now known to occur in various plants and fungi, though it is rare for more than a few to be found in any one organism. The gibberellins are identified by the letters GA and a number. The first gibberellin isolated and identified was called gibberellic acid and it was later given the label GA_3.

Despite the fact that they differ only slightly in chemical structure, plants can easily recognize the various gibberellin molecules. They will react dramatically to one, and yet remain quite unaffected by another. Bean stems, for example, grow strongly in response to treatment with GA_3, only very slightly in response to GA_1, and not at all in response to GA_4. On the other hand, treatment of the plant *Myosotis* with GA_1 promotes flowering, whereas GA_3 and GA_4 are without effect, and in *Silene* only GA_7 promotes flowering while GA_1, GA_3, GA_4 and GA_5 have no effect.

The most characteristic and unique activity of the gibberellins is that they are able to cause dramatic growth in dwarf plants. For example, the growth induced in dwarf pea seedlings by the application of only one-millionth of a gramme of GA_3 to each plant is so great that after a few weeks the plants are indistinguishable from a normal tall variety. But perhaps the most dramatic effect of the gibberellins is to promote internode growth in rosette plants. An internode is the section of stem between the points where the leaves are attached. In rosette plants like lettuce and cabbage, and in extreme examples like the house leek, the internodes are extremely short, yet even one-billionth (thousand millionth) of a gramme of GA_3 will make these internodes grow to such

an extent that a cabbage or lettuce will reach four or five metres in height if supported by an appropriate pole!

Gibberellins can cause flowering in long-day plants even when they are kept in short days, and they can also induce stem elongation and flowering in biennial plants that normally required a period of cold or vernalization (Chapter 15). Furthermore, in Chapter 5 we considered the involvement of gibberellins produced by the embryo in mobilizing the endosperm food reserves in cereal seeds. Gibberellins promote the germination of a wide variety of seeds, some of which are otherwise

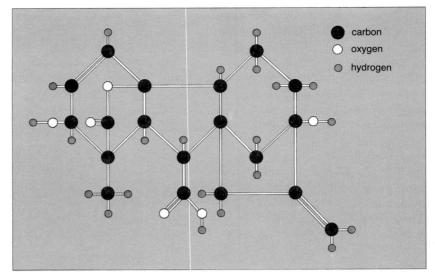

carbon
oxygen
hydrogen

The molecular structure of one of the gibberellins, gibberellic acid (GA_3), is shown here (*top of page*) to give an idea of the molecular complexity of these 'hormones'.

The growth-promoting effect of one-millionth of a gramme of gibberellic acid is shown quite dramatically by the treated and untreated samples of dwarf pea seedlings above. The treated plants are indistinguishable from the normal (tall) varieties of peas.

extremely difficult or slow to germinate, and they can also substitute for the red-light requirement of the 'Grand Rapids' variety of lettuce seeds, causing them to germinate readily even in darkness. Gibberellins therefore seem to be involved in a wide range of important development and growth processes in plants, but except for the case of the aleurone layer cells, we lack a thorough understanding of how they operate.

The gibberellins are now very important compounds in the agricultural and horticultural industries, and also in brewing and malting. They are used to increase the size of seedless grapes, and are applied to celery to produce larger and crisper stalks. In the brewing and malting industries gibberellins are applied to the germinating grain during malting to promote even germination and the early release of sugars from the endosperm tissue for maximum alcohol production. The advantage of their use is that they are wholly naturally-occurring substances.

The cytokinins

The cytokinins are essentially chemicals that regulate the process of cell division, and their chemical structure is quite different from that of auxin or the gibberellins. They all share a common basic structure, and are distinguished one from another by the nature of the side-chain that is attached to the adenine molecule.

The first indication that cell division factors existed came in 1912 when the juice from phloem tissue was shown to cause active cell division in potato slices. Another rich source of a cell division promoter was found to be coconut milk. Identification of the type of chemical structure involved came, however, from a totally unexpected source – the sperm of fish! It was discovered that when herring sperm was heated under pressure to about 120°C, a substance was formed which strongly promoted cell division, especially in cultures of the pith cells from the centre of a tobacco stem. This substance, called kinetin, does not occur naturally in plants, but its identification demonstrated the kind of molecular structure likely to be associated with the naturally occurring cell division factors, a number of which have now been isolated.

Cytokinins are most abundant in growing fruits, seeds and roots. They appear to be synthesized in the roots and transported in the xylem sap to the aerial parts of the plant.

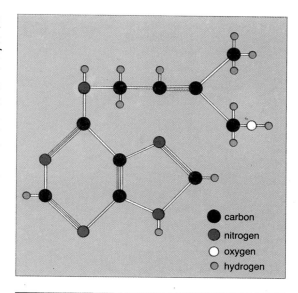

This diagram illustrates the structure of a molecule of zeatin – a naturally occurring cytokinin from a sweetcorn plant.

carbon
nitrogen
oxygen
hydrogen

If a few plant cells are placed in a sterile container on an agar medium containing nutrients and exactly the right amounts of IAA and the cytokinin kinetin, a mass of more or less spherical cells can be grown. There is no organization of the cells into recognizable plant organs such as roots or shoots: the uniform mass of cells is called an undifferentiated callus culture. The ability to grow plant cells like this makes it possible to develop new types of plant by genetic engineering.

However, they are also sometimes combined chemically with a sugar molecule, and in this combined state they move around the plant in the phloem tissue.

Cytokinins have a number of different physiological activities, the most characteristic of which is their capacity to promote cell division. If tobacco pith tissue is placed on a nutrient medium containing only auxin the cells enlarge and the nuclear material may replicate, but *cell* division appears to be inhibited. Addition of a cytokinin causes prolific cell division so that a mass of tissue is obtained. Another very important property of cytokinins is that they interact with IAA to control organ initiation. The relative amounts of IAA and kinetin can determine whether a tissue culture remains undifferentiated, that is a mass of more-or-less spherical cells with no organization at all, or whether those cells differentiate and develop into structures like shoots or roots. By arranging the correct balance of IAA and kinetin in the nutrient medium, the development of *either* roots *or* shoots can be arranged at will.

Cytokinins may also have a role in controlling the senescence of plant organs such as fruits and leaves. The gradual yellowing of a mature leaf in autumn can be arrested by treating it with cytokinins. The effect is to halt the export of proteins, RNA and lipid molecules, and to prevent the breakdown of the green pigment chlorophyll so that the senescence of the leaf is halted. It is possible in a few plants to persuade roots to form at the base of the stalk of a detached leaf and when this happens the leaf can be kept for a year or more in a green, healthy and functional condition.

In answer to the question of how the cytokinins actually work, there is evidence that they bind to active receptor sites in protein molecules on the ribosomes in the cytoplasm. These are the bodies specifically associated with the assembly of amino acids into proteins which have enzyme activity. Much of the activity of the cytokinins is prevented by the presence of chemicals that block protein and messenger RNA synthesis, and many of the processes affected by cytokinins – cell division, reversal of senescence, and lateral bud growth – clearly involve enzyme and protein synthesis.

The cytokinins are a spectacularly active group of chemicals that occur naturally and appear to interact with IAA in regulating organ development besides being essential for cell division. However, our knowledge of just how they work at the molecular and cellular level is virtually non-existent at the present time.

The inhibitors

There occur in various plants a number of chemical substances that appear to inhibit growth, development and metabolism. These compounds, unlike auxin, the gibberellins and the cytokinins, are not a closely related group of substances, but have quite a range of chemical structure.

The most widespread inhibitor of growth and development is a substance called abscisic acid.

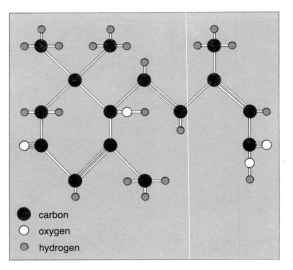

carbon
oxygen
hydrogen

The diagram on the left shows the structure of abscisic acid, a potent growth inhibitor that occurs naturally in plants.

Lettuce seeds soaked in a very dilute solution of abscisic acid for eight days (*left*) show little sign of germination, but removing the abscisic acid and replacing it with water (*below*) leads to perfectly normal germination three days later.

This chemical is physiologically active when applied to plants at extremely low concentrations – at or below one-millionth molar or about 0.2 parts per million! Its effect is generally to 'switch off' almost all metabolic activity in the plant cells, so that everything is either shut down or slowed to a very low rate indeed. But in doing this, the substance does not damage the plant; immediately it is removed, normal growth and development resume. If lettuce seeds are soaked in a dilute solution of abscisic acid and kept warm and in the light with adequate oxygen, they will take up water and swell, but absolutely no germination takes place. The seeds will remain in this state of 'suspended animation' for several weeks, but if they are then transferred to water they begin to grow and develop into normal lettuce plants.

Dormant winter buds of many plants contain quite high levels of abscisic acid, and the amount decreases as the buds begin to break in the spring. Similarly, some very dormant seeds contain quite large amounts of abscisic acid, and they do not germinate until the inhibitor has been washed away by the soil water. However, the extent to which abscisic acid is the regulating chemical in controlling the onset and breaking of dormancy has not really been established.

Many scientists now believe that the principal role of abscisic acid in a plant is in its water relations, particularly in helping the plant conserve water in times of drought. Many herbaceous plants show a dramatic rise in their abscisic acid content when they are exposed to a dry atmosphere, or indeed subjected to any treatment that causes them to wilt. Although there are some differences between different species the loss of the cells' hydraulic pressure

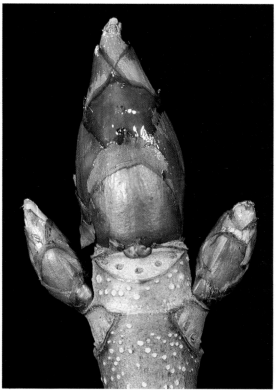

Winter buds actually form in the summer, long before the cold winter weather arrives. They can be found nestling among the foliage leaves, as in the horse chestnut branch above, photographed in August.

Winter buds consist of a large number of scale leaves which completely cover the growing tips of the stem and lateral branches. They are often waxy or resinous for added protection. Those of the horse chestnut (*far left*) are particularly sticky.

Bud break in spring seems to be triggered by a change in the plant's hormone balance, notably by a decrease in the levels of growth inhibitors such as abscisic acid. In this photograph of a sycamore bud (*left*) well-developed young foliage leaves are pushing aside the scale leaves of the winter bud.

appears to coincide with, and may therefore trigger, the rise in abscisic acid level. One of the things that abscisic acid does to the surface of leaves is to cause the guard cells surrounding the pore to lose their hydraulic pressure and collapse – thus closing the pore.

Despite a great deal of investigation no one has yet identified the characteristics of the abscisic acid molecule that are essential for its activity. Neither has there been any satisfactory evidence for a site of action for abscisic acid; indeed it is likely that the molecule acts at a number of different sites because of its rapid effects on membranes and long-term repression of growth and development.

Abscisic acid is quite expensive to make, and this alone has largely prohibited its use in commercial horticulture. Another problem is that it has not really been found to have a useful role; no one wants plants with winter buds or their growth totally inhibited! Nevertheless, there is considerable commercial potential in this substance because plants sprayed with abscisic acid are more resistant to drought, and also apparently to certain types of injury such as chilling. To be able to increase a plant's ability to withstand sudden water stress would be of considerable value in countries where crops are frequently subject to drought.

Ethylene

If readers suppose for a moment that they are in the happy position of being The Creator, with the task of designing and building a green plant, it is unlikely that they would even consider incorporating a regulating system involving a gas. Such an idea would probably be rejected for no other reason than that every puff of wind would change the concentration and distribution of the gas with the result that growth and development would be continuously upset. But incredible as it may seem, a gas does appear to operate as a hormone in almost all plants, the gas in question being ethylene (C_2H_4). Ethylene occurred to a small extent as a constituent in town gas in Victorian England, and as early as the turn of the present century it was known to cause deformation, leaf senescence and death in most house plants of the time. Indeed, the popularity of the aspidistra as a Victorian house plant rested solely on it being largely unaffected by ethylene and therefore able to tolerate the atmosphere in a room in which gas was being used for both lighting and heating.

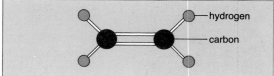

The popularity of the aspidistra as a house plant in Victorian England was due at least in part to the fact that it was one of the few plants that could tolerate the levels of ethylene gas produced by indoor town gas lighting.

Ethylene is a relatively simple molecule consisting of two carbon atoms and four hydrogen atoms (*left*).

Ethylene affects so many aspects of plant growth and development that one begins to wonder if it is not a kind of general poison that upsets the natural operation of plant cells in such a way as to give rise to all kinds of abnormal behaviour patterns. However, current thinking does not believe this is so, but rather that it is an important natural regulator of growth and development in plants.

When present in the atmosphere at concentrations as low as 0.16 parts per million, ethyl-

Ethylene is an important regulator of the ripening process in many fruits, and as such is of great value in controlling the ripening of fresh fruit in transit to markets around the world.

ene causes gross deformity of the stems of dark-grown pea seedlings, which become thick and swollen, and bend over to grow horizontally. Ethylene also promotes the shedding or abscission of leaves, breaks dormancy in buds and seeds, releases apical dominance, promotes the ripening of many fruits, initiates flower development and can even change the sex of flowers that normally develop only male or female organs!

Ethylene is actually produced by plants, of that there is no doubt. All parts of the plant appear to be capable of producing the gas but the most productive parts are those where a lot of cell division and growth is occurring, for example, at the apex of the stem and roots. When a seed germinates, the highest rate of ethylene production occurs as the root forces its way through the seed coat. In flowers, the highest rate is when they begin to fade. Fruits also produce large amounts of ethylene, especially apples, bananas, avacados and tomatoes, which are called climacteric fruits because in their ripening process they show a marked burst of CO_2 production and at the same time a dramatic rise in their ethylene production. It is this ethylene production, especially by a rotten apple in store, that will cause all the other apples nearby to ripen more quickly and hence spoil. Similarly an orange will produce ethylene, which may be used to make bananas ripen much more quickly than if the orange is not present. In storing bananas and transporting them, ships, containers and trucks must have excellent ventilation systems so that the ethylene produced by the fruit does not accumulate: if it does, it will so hasten the ripening process that the fruit will be ruined before it can be delivered to the shops.

Two substances, carbon dioxide and silver nitrate, have been found to interfere with the activity of ethylene, and both are used commercially. Carbon dioxide is used to prevent the ripening of fruit in storage, while silver nitrate is sprayed on to ornamental foliage and cut flowers to prevent ethylene production and so extend their shelf-life. This treatment is particularly effective in cut carnations.

Ethylene has become one of the most important of all plant hormones from the commercial point of view. A ready source of the gas is the chemical called 2-chloroethyl-phosphoric acid, sold under the trade name 'Ethrel'. This substance releases ethylene and is used as a means of influencing the ripening of tomatoes, apples and grapes, and for making easier the harvesting of cherries, blackberries, walnuts, grapes, blueberries and cotton by promoting the abscission of these fruits. 'Ethrel' also assists in prolonging the flow of latex in rubber plants, and in increasing the sugar content of sugar cane plants. It is also used to promote the flowering of pineapples and in accelerating the senescence of the leaves of tobacco plants. Knowledge of the effects of ethylene in promoting fruit ripening is now used to design and operate modern storage and transport facilities for fruits so that they can be sold to worldwide markets, while inhibitors of ethylene action can greatly improve the shelf-life of cut flowers.

CHAPTER 17

Living Together
Symbiosis and Scrounging

There are many instances of plants living together in apparently harmonious and mutually advantageous relationships. Indeed, we have already met two such cases in Chapter 11. In the first, the bacterium *Rhizobium* and various leguminous plants such as peas, beans and clover enter into a relationship that produces root nodules with the capacity to fix atmospheric nitrogen. In the second, the roots of forest trees and orchids form a special relationship with fungi to form mycorrhizas. These structures considerably enhance the ability of the green plant to absorb mineral nutrients while simultaneously providing the fungus with a supply of carbohydrates. Such relationships are called mutualistic symbiosis.

There are, however, many other cases of plants living together in which the relationship is definitely not to the benefit of both partners.

One is quite clearly living at the expense of the other, in some cases to such an extent that the provider becomes sick, weak and eventually dies. This type of relationship is termed parasitism, and is characterized by one organism, the parasite, living at the expense of the other, the host. In reality, a parasitic relationship is just an extreme, and one-sided, form of symbiosis, and the phenomenon is sometimes referred to as antagonistic symbiosis.

In parasitic relationships we meet all the members of the plant kingdom who can not or will not support themselves. These are the scroungers. But worse than that, we also meet the criminal classes of the plant kingdom – the thieves, robbers and murderers who, in order to survive and obtain their supply of nutrients, will steal from, bleed to death and kill other plants. Such organisms include many of the

Like all lichens, *Cladonia macilenta* consists of two quite distinct and separate plants, one an alga and the other a fungus. The two live together in a mutually beneficial relationship.

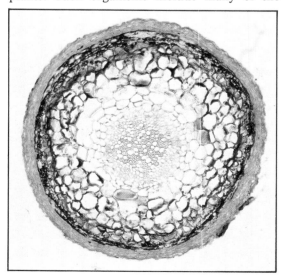

A mycorrhizal association is one in which a higher plant and a fungus live together to the benefit of both partners. The short lateral roots of the plant become totally enclosed in sheaths of fungal hyphae, many of which can be seen extending over the surface of the root in the far left photograph. In the magnified cross-section (*left*) the sheath of pink-stained hyphae can be seen covering the whole root surface.

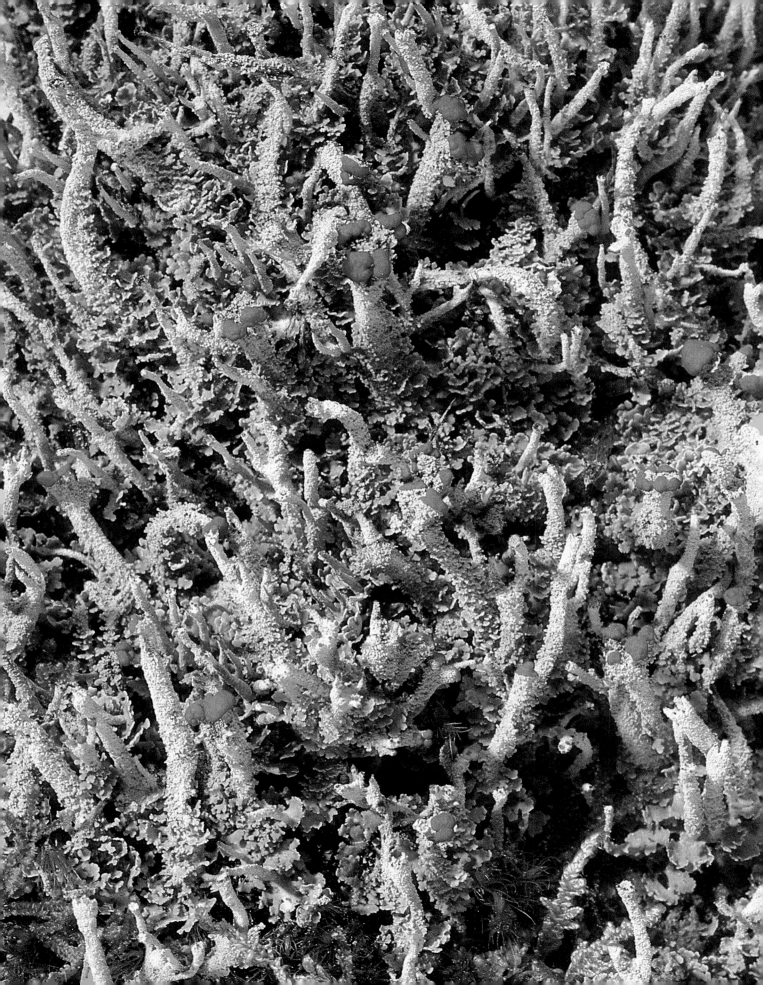

bacteria and fungi, all of which have to obtain their supply of carbohydrate nutrients from somewhere because they lack chlorophyll and therefore can not make their own using the sun's energy. Many fungi obtain all their carbohydrate nutrients from dead organic matter, plant and animal remains, in the soil. They are called saprophytes. Others derive their nutrient supplies from living plant and animal tissues, and these are the parasites. Many are relatively harmless, but others are quite lethal to their hosts and include the principal pathogenic or disease-causing organisms that can wipe out entire crops and cause serious illness and death to animals and man. Fungi are totally unscrupulous in their criminal activities: they even parasitize other fungi – an example of cannibalism at the microscopic level!

But not all plant parasites are fungi: many flowering plants, too, have adopted the way of life of the thief or robber by attaching themselves to, and deriving their nutrients from, some other independent green plant. These parasitic flowering plants have little or no alternative, because they are all either very deficient, or totally lacking, in chlorophyll and so can not support themselves. Presumably this loss of chlorophyll occurred in a genetic mutation, and the only way for the plants to survive after that was by adopting a parasitic way of life. Nevertheless, it has proved a successful way of life because parasitic plants survive in quite large numbers wherever suitable hosts grow.

Many plant relationships are not very precise or specific, but still involve a degree of dependency. The trees in a forest, for example, provide the dim light and humid environment that shade-loving plants like ferns require. Many larger plants provide rigid structures up which other plants can climb, and on whose surfaces they can attach themselves and grow without ever making contact with the soil beneath. Plants that grow high up in trees and are merely attached to them without deriving any nutrients from them are called epiphytes. Epiphytic plants are found on most trees. Mosses and lichens are good examples in temperate regions, while in tropical rainforests many of the orchids and bromeliads (members of the pineapple family) are entirely epiphytic. These relationships are really the province of ecology, but they are mentioned here for the sake of completeness. This chapter is really concerned with those cases in which plants enter into

very close relationships involving cell to cell contact, and where there is either a mutually beneficial exchange of nutrients or a situation in which one partner partially or completely nourishes the other.

In discussing these relationships we often have to refer to their 'specificity', which simply means that one plant species is very particular about the species of the partner with which it will form a relationship. Sometimes even a particular strain within a species is required – all other strains being rejected. This specificity in setting up a relationship has been most closely studied in the case of the nodule-forming bacterium *Rhizobium* and its leguminous hosts. In a particular species of soya bean, for example, only certain strains of the bacterium will enter the host roots and initiate nodule formation; other strains have no effect. If the soya bean seeds were to be sown in soil containing only these non-active *Rhizobium* strains then a disastrous crop would result – hence the economic importance of understanding such processes. From the scientific point of view the main question of interest in such studies is that of exactly how the soya bean root recognizes *Rhizobium* cells of a particular strain, and sets up a nitrogen-fixing relationship with them, while totally ignoring other strains. These other strains are often active in forming nodules in other plants, so it is not a case of them being unable to induce nodule development. The answer lies in a complex and very sophisticated molecular recognition system that operates at the level of the individual cell wherever two organisms come into direct contact.

Rafflesia produces one of the largest flowers in the plant kingdom. It is a root parasite, and except for its enormous flower, grows almost entirely within the tissues of the host plant's roots.

Many different kinds of relationship can be established between plants. Usually, for example, there are two partners, but there are cases in which three different plants set up house together – not usually a formula for success from human experience! The partners in the more normal dual relationships can be two flowering plants, a flowering plant with an alga, fungus, or bacterium; an alga and a fungus; a fern and an alga; two algae, or even two fungi.

The least specific relationships are those in which one plant merely uses the other for support. Most of the climbing plants that are independently rooted in the soil and have photosynthetic leaves fall into this category, whether they attach themselves to the supporting plant by twining tendrils, hooks, or by special adhesive organs. Ivy is an independently rooted climber that attaches itself to the surface of a tree or wall by specialized adhesive adventitious roots which grow out from the main stem. The ivy and its host are obviously very firmly stuck together, but there is no evidence at all that even the adhesive roots draw any nourishment from the host. The same is true for epiphytic plants, but these of course are not independently rooted in the soil. The epiphytic orchids in tropical rainforests are usually located in the angles of the branches of the trees, or attached to other crevices in the trunks. Many produce enormously long roots which hang down in the humid forest atmosphere, rarely reaching the ground. These aerial roots have a very interesting structure and are believed to absorb at least some of the water required by the orchid. The rest is probably obtained from the small pools, and clumps of soggy plant debris, that collect in the angles between the branches and trunks of large trees. In the case of the epiphytic bromeliads, the leaves are arranged in an inverted cone-shaped structure which collects and holds water. Epiphytic ferns and mosses appear to subsist on rain-water that trickles down the tree trunks.

The main problem for epiphytic plants is to obtain an adequate supply of mineral nutrients, and these appear to be collected from trickling rain-water, the ion content of which is no doubt enriched by the leakage of ions from the dead and dying cells of the host tree's bark. Epiphytic plants do not appear to derive any other nutrients from their hosts. They all have photosynthetic leaves, and none of them make any attempt to penetrate the host plant's tissues

The green aerial roots of the epiphytic orchids are actively photosynthetic, although the green tissue is often hidden by an extra layer of silvery cells called a velamen (*left*), which covers all but the tip of the root. A cross-section (*below*) shows that the velamen is separated from the cortex and central stele by a layer of large cells, the exodermis, shown again at higher magnification in the bottom photograph. The velamen is thought to be associated with the absorption of water from the damp forest atmosphere.

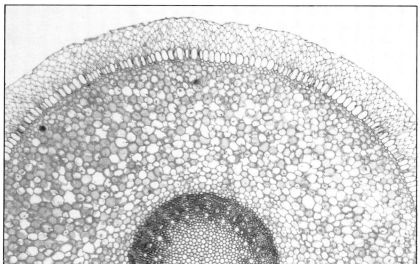

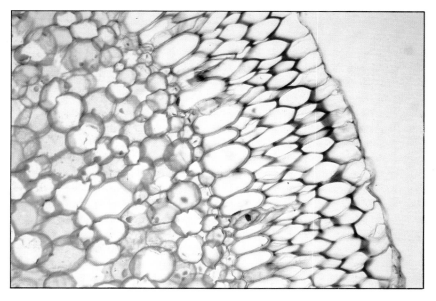

in any way. Epiphytes, then, do little more than 'hitch a ride' on the host plant, and their principal gain is simply a more favourable position, higher in the sun-lit layers of the canopy.

On the other hand, many of the plants that grow on other plants depend on their host for some or all of their nutrients, and because of this requirement the host plant's tissues are usually penetrated by organs of the passenger plant. The living tissues of the two plants thus grow into close contact so that nutrient transfer can take place. If the passenger plant derives only a part of its nutrient requirement from the host it is known as a partial parasite or hemi-parasite, and such parasites often attach themselves to the stems or roots of their hosts. A very common stem hemi-parasite is the mistletoe, which grows high up on the branches of, for example, apple trees. The bark of the tree is penetrated by a specialized root-like structure which eventually reaches the secondary xylem of the branch and spreads out to form a root-like absorbing organ. In this way the mistletoe plant obtains its water and mineral ions from the host, but it provides its own carbohydrate nutrients through the photosynthetic activity of its own prolific green leaves. Other hemi-parasites attach themselves to the roots of their host, as in the case of the witchweed, the roots of which penetrate those of such important crop plants as maize and sorghum causing a substantial reduction in yield. There is often a high level of specificity between parasite and host plants in these hemi-parasitic relationships, as might be expected where living cells come into contact to facilitate nutrient exchange.

Many other plants are, however, totally parasitic, drawing their entire nutrient supply from other plants and being incapable of any independent existence except as a seed. Again, because of the intimate contact between the living tissues of the two partners there is usually a high level of specificity. A number of flowering plants are totally parasitic because they have lost the capacity to synthesize chlorophyll. The most familiar plants in this category are the dodders and broomrapes. The dodder is a stem and leaf parasite that twines itself around and all over its host, its own extremely long and thin branched stem making frequent close contact with the host's stem. At these points, lateral root-like organs penetrate the host stem, quickly developing xylem and

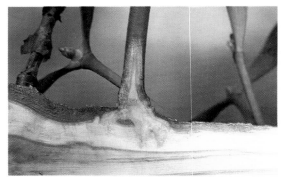

The familiar mistletoe is a partial parasite. It carries out photosynthesis in its own abundant green leaves (*above*) but obtains all its water and mineral requirements by driving highly specialized root-like structures into its host to tap the host plant's xylem and phloem. The section (*left*) shows a branch of an apple tree, penetrated by mistletoe.

phloem transport tissues that join up with the vascular tissues of the host. Particular species of dodder are often very specific about their choice of host; *Cuscuta europaea*, for example, is usually found only on nettles and very rarely on other plants, while *Cuscuta epilinum* seems to grow only on flax.

The broomrapes are root parasites. The seeds seem only to germinate when in contact with the roots of an appropriate host, and the emerging seedling's root immediately penetrates the host root tissues, eventually establishing a connection between the nutrient-carrying vascular tissues of the two plants. Later, the flowering shoot, brown in colour and totally devoid of chlorophyll, emerges above the ground. The broomrapes also have a fairly high degree of specificity in their host range. The purple broomrape, for example, grows only on roots of *Achillea millefilium* and some other members of the Compositae family. In contrast, the toothwort appears to be much less particular about its relationships since many woody plants of woodlands and hedgerows appear to satisfy its requirements.

The broomrapes (*Orobanche, right*) derive all their nutrients from the roots of their host plants. They are completely devoid of chlorophyll and so have no green parts. Their small scale-leaves are brown and their flowers a purple-brown.

The toothwort (*Lathraea squamaria, far right*) is also a root parasite, attaching itself to, and deriving all its sustenance from, the roots of the host plant. The toothwort produces spectacular pinkish-white flowers.

In some rather more complicated partnerships between higher plants, a third organism is used to link the two main partners together. The dicotyledonous plant *Monotropa uniflore*, the Indian pipe, for example, is totally devoid of chlorophyll and sends up completely white flowering shoots with minute white leaves. It was believed for many years that this plant was a saprophyte rather than a parasite, but it is now known that *Monotropa* can only exist if it is associated with a fungus – which, of course, also lacks the capacity for photosynthesis! This fungus, however, is one that forms a mycorrhizal association with other green plants such as spruce and pine trees, and part of the carbohydrate exported from the green plant is transferred by the fungus to the *Monotropa* plants, along with mineral ions and other nutrients!

Many parasitic plants have become so modified by their nutritional requirements that they have almost completely dispensed with a vegetative body and instead grow almost entirely within their hosts' tissues, sending out their extraordinary flowers from time to time. Many members of the family Rafflesiaceae fall in to this category. In one, *Pilostyles*, which parasitizes some asiatic species of *Astragalus*, the plant's body appears to consist of a series of single-celled filaments which penetrate the host tissue like the hyphae of a fungus. The flowers are the only part of the parasite that appear on the surface of the host, and these appear, from time to time, at the base of the host plant's leaf stalks.

Surprisingly, in the relationship between a flowering plant and a fungus, it is not always the fungus that is the villain. A number of flowering plants are, in fact, parasitic on fungi. The birds'-nest orchid, for example, is quite devoid of chlorophyll and can not therefore synthesize its own carbohydrates. Its roots are thick and short, and form a dense clump, and this hardly makes them efficient at absorbing mineral ions from the soil. How then does this plant acquire enough essential nutrients to build up its elaborate body? The outer cortex of the root is infested with fungal threads, which penetrate the living cells and appear there to undergo digestion. This is in effect an endotropic mycorrhiza, because the fungal hyphae penetrate the soil where they break down organic matter to acquire a carbohydrate supply as well as absorbing mineral nutrients. These are then transferred to the orchid which, in effect, is parasitic on a saprophytic fungus!

Many other mutually beneficial relationships exist between green plants and fungi, a spectacular example being the lichens, a group of so-called plants which are not really a separate group at all because a lichen is really an alga and a fungus living together, each being clearly recognizable as a member of its own distinct group. More than 90 per cent of all lichens are formed by the association of a member of the group of fungi called ascomycetes with one of two species of unicellular green algae called *Trebouxia* and *Trentepohlia*, or with the photosynthetic cyanobacterium *Nostoc*. The association between the two organisms produces the lichen, whose structure can vary enormously, from the flat silvery thallus of *Parmelia*, through brightly coloured powdery deposits on rocks and orange-yellow encrustations on tree trunks to the massive green lichens like old man's beard which festoons tree branches.

One remarkable consequence of the association of a fungus and an alga to form a lichen is that the resulting structure is able to withstand an astonishing range of climatic and environmental conditions. Lichens are found virtually totally dried up in desert areas where their water content may be as low as two per cent. On wetting, however, they quickly absorb large amounts of water, the cells become metabolically active, the algal cells start to photosynthesize and the whole organism springs into life. As the structure dries out again its metabolism slows, and when a state of extreme desiccation is reached the lichen goes into 'suspended animation' and remains in that condition until the next rainstorm. Lichens occur not only in the heat and dryness of desert areas, but also in extremely high and cold areas such as alpine peaks and polar regions. There is even a marine form – the frequently submerged organism *Verrucaria serpuloides*. The reindeer moss is of critical economic importance to the Laplanders in Scandinavia and the Eskimos in Alaska because it is the principal food of reindeer and caribou, the nomads' chief source of food, clothing and other essential materials.

As is evident from their bright colours, lichens can synthesize a wide variety of pigments, and they have been found to contain a

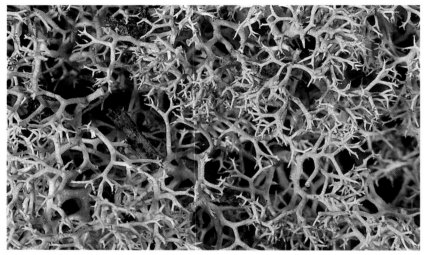

considerable number of quite unusual chemical compounds. They grow extremely slowly and a very large lichen may be as much as 4,500 years old. They reproduce vegetatively either by parts of the plant breaking off and beginning an independent existence, or by the production of minute bodies called soredia which consist of a few algal cells to which are attached some small pieces of fungal hyphae. These small bodies are dispersed by wind and rain to generate a new lichen plant where they alight.

So what does each partner give to, and take from, this association? Certainly the lichen is able to exist in environments that would be intolerable to an alga alone, and that would probably also be intolerable to the fungus (although few, if any, of the fungi involved seem able to exist on their own anyway). The algae and the cyanobacteria involved in the formation of lichens can all live alone naturally and successfully, and they do so in large numbers. Some actually grow more rapidly when

Lichens are extremely good indicators of environmental pollution, and soon disappear from areas where the air is contaminated. The branching leafy and shrubby varieties are particularly sensitive to the sulphur in exhaust gases and industrial smoke.

At least four different lichens can be identified in the top photograph – a dead tree branch lying on the ground in the highlands of Scotland. The reindeer moss *Cladonia rangiferina* (*above*) is the principal food of the reindeer of the Arctic.

free-living. The fungi, on the other hand, are hardly ever found by themselves in nature, although they can be grown readily in the laboratory. Experiments have shown, however, that these fungi require a complicated diet of nutrients if they are to grow satisfactorily. Studies using radioactively-labelled sugars have shown that in the lichen association, sugar molecules like glucose move from the algal cells to the fungus. When the algal member of the association is the cyanobacteria *Nostoc*, an organism that can fix atmospheric nitrogen as well as being photosynthetic, then the fungus will gain nitrogenous compounds as well as sugars from its partner. The algal cells appear, however, to be affected by the presence of the fungi because there is evidence that their sugar metabolism is changed considerably.

Since the algae live without the fungi, but the fungi can not apparently live without the algae, at least in nature, it may be that a lichen does not represent a real mutualistic symbiotic relationship, but rather a parasitic one in which the algal cells are parasitized by the fungus and are more or less enslaved to provide the latter with its specific nutrient requirements. The fungal hyphae certainly surround and ensnare the algal cells, and attach to them short lateral branch structures which presumably act to absorb from the algal cell the leaking carbohydrate nutrients.

The nutrients required by lichens are absorbed from the surface to which they are attached, or from the minerals inevitably present in rain-water. Like all plant cells, the lichen partners absorb mineral ions rapidly when they are wet, and this leads to a considerable concentration of the mineral ions in the lichen itself. This has had disastrous consequences for those animals, and people, who depend on lichens as the first stage in their food chains. Radioactive fall-out from the testing of atomic bombs, and most recently from the disaster at the Chernobyl nuclear power station near Kiev in the

USSR, has resulted in the reindeer moss lichen in Lapland building up huge concentrations of radioactive substances. The contaminated lichens have been eaten by the reindeer, which in turn have incorporated the radioactive substances into their bodies. At the time of writing, the Lapland reindeer herds cannot be culled for food because the meat has been declared too radioactive for human consumption. Furthermore, the radioactivity in the reindeer moss will have to decay naturally before reindeer meat will once again be safe to eat – and, unfortunately, not only does the lichen tend to be a long-lived plant but the accumulated radioactive ions are also very slow to decay. Clearly, we are going to have to be much more careful about the way in which we manage the environment, otherwise in the not too distant future our planet may once again find itself populated only by plants.

The deadly invaders

The most extreme cases of plant interdependence are those that are disastrous for one partner and beneficial only to the other. These are the pathogenic associations caused by disease-inducing organisms, usually bacteria, fungi and viruses. It is obviously not in the interest of pathogenic organisms to kill their host too quickly, or even at all, and certainly not before they can grow and produce large numbers of reproductive bodies of one sort or another which can infect other plants. But many infections of plants by micro-organisms are lethal, and their economic consequences are horrendous. Disease-causing agents have led to untold misery, starvation and death throughout the world. The infection of the potato plant in Ireland by the fungus *Phytophthora infestans* has, on several occasions in the past, wiped out the entire crop, leading to widespread suffering and large-scale emigration of people to mainland Britain, America and Australia.

The diseases caused by bacteria are very varied, and a great many of them are extremely damaging to the host plant. Even if an infected crop plant survives to maturity, the yield is usually low and of very poor quality. Amongst the most important bacterial infections are the cancerous growths induced by *Agrobacterium*, the ring-rot of potato and tomato wilt diseases caused by *Corynebacteria*, various wilts caused by *Erwinia*, and spot and root diseases often caused by *Xanthomonas*. The blight diseases are characterized by invasion of the leaf and

stem cells by the bacteria and the rapid loss of chlorophyll leading to yellow or black areas of dead cells called necroses. Wilt diseases often involve the bacterial pathogen attacking the vessels of the xylem of herbaceous plants. The bacteria grow and multiply in those important water-transporting cells and essentially block them up so that the upper parts of the plant are deprived of their water supply. Later the bacteria can actually attack and destroy the walls of the xylem cells and the surrounding cells so that the stem loses its strength and collapses.

Fungi, too, cause a wide range of plant diseases. Many attack stored fruit, causing large-scale economic loss, while others cause rot and damping-off diseases of young seedlings. In all such diseases the infecting fungi induce a rapid break-down of the host tissues and the rapid death of the seedling or plant. Specificity is not very high in these cases. A number of pathogenic fungi have a less serious effect on the host plant, but even though they do not kill it, they do seriously reduce its crop yield. Many of these fungi, especially those causing the rust and smut diseases of cereals and other plants, are of economic importance and there is usually a highly specific relationship – each species or even strain of fungus attacking only a particular species or variety of host plant.

The high degree of specificity characteristic

The *Penicillium* fungus above will attack and grow on a wide variety of materials, in this case an orange which is clearly ruined. The dull greenish patch is a region of spore production. This fungus is closely related to the one that led to Alexander Fleming's discovery of the antibiotic penicillin.

A crop of wheat, heavily infected with the yellow rust fungus (*Puccinia striiformis*). Note the extensive damage to the leaves. This greatly reduces the plant's ability to photosynthesize, and so reduces its yield.

A sickening sight to any potato farmer: this tuber shows the symptoms of the potato gangrene disease caused by the fungus *Phoma foveata*. Infected vegetables are rendered quite useless.

of many plant relationships clearly indicates that the partners can identify one another precisely and react differently to, or reject, all others. Such identification involves the living cells of the two partners coming into close proximity, or even touching, which in turn suggests that cells of a particular species or strain carry some identifying chemical signal which can be recognized and acted upon by other plants. It is important that these recognition systems are fully understood because they determine which varieties of a green plant are resistant to pathogenetic attack by another plant and which are susceptible. The breeding programmes that have led to the development of resistant strains of crop plants have evidently successfully inserted into the host plant the genes that enable the plant to recognize a particular pathogenic organism and then either limit the spread of the invader in its tissues, or eliminate it altogether. This, of course, is a never-ending battle for the plant breeder because if the pathogenic organism undergoes a mutation, or spontaneous genetic change, that alters its identification signal, then the host plant will again be susceptible to infection because it will no longer recognize the pathogen as dangerous. Such mutations are constantly occurring, and so the need for newly bred resistant varieties of crop plants continues. Conventional plant breeding is successful but inevitably slow, and the possibilities of transferring specific resistance-giving genes from one plant to another by genetic engineering (Chapter 18) are now being explored. These new techniques may offer outstanding and exciting opportunities for the development of disease-resistant varieties of crop plants in the years ahead.

CHAPTER 18

New Plants From Old

In the earliest times man was nomadic and gathered his food as and where he found it, but as soon as he realized that seeds could be gathered and sown in suitable soil to provide crops, he began to live a more settled existence. With the advent of this first, simple, agricultural practice, man began to exert an influence on the development and evolution of plants. By selecting seeds from their biggest and highest-yielding crop plants to sow the following year, the early farmers were propagating the particular strains of plant that best met their needs. Only the most healthy plants would be selected, which means that those most susceptible to disease were discarded. By constantly selecting plants in this way, and allowing them to breed and cross with one another naturally in the fields, even the most primitive farmers were beginning to select and design plants to meet their specific nutritional requirements.

Much later, when an understanding of genetics and sexual reproduction had been acquired, there appeared on the scene professional plant breeders who were able to identify particular characteristics in a plant variety or species and breed them into other strains or species. They were also able to breed out undesirable characteristics. This process of plant breeding involves the very careful fertilization of flowers with pollen specially collected from other plants, and the raising of very large numbers of seedlings which are carefully selected for the desired characteristics and then propagated to provide enough seed for commercial use. It is a slow and laborious process, but one that has been spectacularly successful.

The contribution made by plant breeders to the relief of hunger and starvation in the world stands out as one of the greatest contributions of all time to the welfare of mankind. All the advances in medicine and technology are of little use to a man, woman or child who is dying of starvation, and the transformation of crop production during the past 25 years in the 'green revolution' has succeeded beyond anyone's wildest dreams. Of particular importance have been new varieties of wheat and rice which have led to countries like India, for example, producing grain surpluses where before there were only deficits. The development of present-day potato varieties from the primitive *Solanum* species of South America is evidence of the way better crop yields and the production of larger and better tubers have been achieved. In the same way, the minute fruits of the wild strawberry plant are dwarfed by those of the commercial varieties of today, while the size of the ear on primitive, unimproved wheat plants bears no comparison with the size of the ear on present-day commercial varieties which are also much more resistant to diseases caused by fungi.

Of course, there have been major contributions from many other areas of plant biology as well. Understanding the mineral and water requirements of plants has paved the way for many improvements in agricultural practices, while the development of new pesticides and selective herbicides has made an enormous contribution to the improvement of crop yields. Climate, of course, is a major limiting factor, and not all countries are fortunate in this respect: many are subject to unpredictable and prolonged periods of drought. Provision of water for agriculture in these arid regions is a complex and difficult task because the constant

The glowing tobacco plant opposite is a spectacular demonstration of plant genetic engineering. The plant has been transformed by the insertion into its chromosomes, from those of a firefly, of the gene for synthesizing the enzyme luciferase, which breaks down a substance called luciferin and releases photons of light in the process. When the plant was later watered with a solution of luciferin, it glowed with enough light to take its own photograph, proving beyond doubt that it contains, and can use, the implanted gene.

Plant breeders have been very successful over the years in developing wheat varieties with shorter and shorter stems. This helps protect the crop from wind and rain damage, produces less unwanted straw, and ensures that as much as possible of the captured solar energy goes into producing the grain.

The heads of wheat (*below, left*), oats (*centre*) and barley (*right*) provide much of the food required by man and domestic livestock. In the case of barley, the plant also provides sugar for fermentation in the beer and whisky industries.

high levels of evaporation of irrigation water create heavy concentrations of salt in the surface layers of the soil – and most plants will not tolerate saline conditions. Eventually, however, it is hoped that plant breeders and genetic engineers will be able to increase the salt tolerance of some important crop plants, perhaps by introducing appropriate genes from naturally salt-tolerant strains. There are, for example, wild varieties of tomato that live in coastal habitats and show a high degree of tolerance to salt spray.

In the breeding of new crop plants, a major objective is to make the plant channel more of the energy it captures from the sun into the parts we can use as food, and less into those parts that are of little or no value as food. The modern varieties of wheat, barley and oats provide excellent examples of such changes. At the turn of the century, most cereals had extremely long stems – wheat and oats often being up to 1.5 metres tall. Their height made the plants very vulnerable to being knocked down by wind and rain, and a laid crop was difficult to harvest. Breeders have succeeded in reducing the length of the stems of most cereals so that barley and wheat, for example, are now seldom more than 65 centimetres tall. Much

more of the captured solar energy now goes into the edible grains, and much less into the unwanted and rather useless stem.

Cross breeding between different species of a particular plant genus usually gives rise to a plant that is more vigorous in its growth than either of the two parents. No one quite understands the reason for this hybrid vigour but it is very important in present-day agriculture since most of the high-yielding varieties of crop plants are hybrids. Unfortunately, hybrids are often sterile, that is, unable to produce seed, while even in those that are fertile the offspring are seldom all of the same type, many being 'throwbacks' to the original parental types. As a result, new hybrid seed has to be produced each year from the two original parental lines. Many species of plant will not, however, interbreed, so in these cases the advantages of hybridization are not available. Using natural breeding methods there are, therefore, severe practical difficulties in transferring a desirable characteristic from one plant to another, and this has led to the exploration of entirely new methods of transferring genetic material.

These new methods exploit recent developments in molecular and cellular biology, and their potential is very substantial indeed. For example, it is not difficult to appreciate the tremendous benefits that would be felt in areas where farmers can not afford nitrogenous fertilizers, if the capacity to form nitrogen-fixing root nodules could be transferred from pea or bean plants to cereals such as wheat and maize. Instantly, productivity might be expected to increase many times over, although it must be appreciated that a proportion of the captured solar energy would be utilized in driving the nitrogen fixing process. But while this is a prime and laudable objective, nitrogen fixation by the bacterium *Rhizobium* is a complicated process involving several genes in the bean and pea plants as well as in the bacterium, and a great deal more research time and effort will be required before this dramatic development is likely to become a reality.

There are, nevertheless, a number of more modest, but no less important objectives which could be achieved by plant breeders and by scientists utilizing the techniques of plant molecular biology. Indeed, some have already met with success. It is, for example, important to increase the amount and variety of amino acids in maize and other cereal grains, since many of the nutritional diseases, such as kwashiorkor,

suffered by children whose diet consists only of maize meal, could be alleviated if the meal contained a higher proportion of certain essential amino acids. Maize meal contains only about ten per cent protein and is particularly deficient in an amino acid called lycine, and also in another one called tryptophan. Plant breeders have been able to develop new varieties of corn containing larger amounts of these two amino acids, but varieties with still higher levels are urgently required. Soya beans, on the other hand, have a protein content of about 40 per cent – a very high value indeed – and weight for weight they have more than twice as much lycine as maize meal. But the amounts of an important amino acid called methionine tends to be low in bean seeds so there is considerable scope for improvement in the legumes as well.

Other major objectives of the genetic modification of plants are to increase their resistance to diseases and their tolerance of drought, cold and salinity, and to improve the flavour, nutritional value and storage life of the fruits, seeds, roots and other plant organs we consume. It might also be possible to develop varieties able to tolerate periodic flooding, and to transfer the nitrogen fixing capacity of legumes to other species. There also exists the possibility of increasing the efficiency of photosynthesis in C-3 plants by abolishing the very wasteful process of photo-respiration (Chapter 9) through removal of the appropriate genes. Many of these objectives are beyond the scope of conventional plant breeding and so it will be necessary to use the techniques of genetic engineering to design or 'tailor' the plants to mankind's particular requirements.

Gene transfer by molecular techniques rather than by normal sexual reproduction has already been achieved in a number of plants. For example, the small sub-unit of the enzyme ribulose bisphosphate carboxylase, responsible for the initial fixation of carbon dioxide in C-3 type photosynthesis, has been transferred from soya bean plants to tobacco plants, and it has also proved possible to transfer genes that effect herbicide resistance into plants of oil-seed rape. But the most spectacular transfer of all so far has been the introduction of the gene responsible for the synthesis of luciferase into plants of tobacco. Luciferase is the enzyme that causes luminescence in fire-flies, and in algae like *Gonyaulax polyedra* and *Nostoc*. It causes the breakdown of the chemical luciferin, releasing energy as photons of light. The transformed plants were found to glow or luminesce clearly in the dark when they were supplied with luciferin, so they obviously possessed the necessary enzyme – and the enzyme could only have been synthesized if the gene bearing the appropriate information was present in the plant's chromosomes. Of course, no-one really wants a luminescent plant, but these exciting results have clearly and imaginatively demonstrated the potential of the techniques. Little has been achieved so far in improving the quality of the nutrition provided by individual crop plants, but there can be no doubt that we are on the brink of achieving major advances in this area of plant biological research.

The basic technique of plant genetic engineering involves the insertion into a plant cell of a length of chromosome from another cell, the length of chromosome containing the gene or genes that determine the particular characteristic that it is desired to transfer into the new plant. This piece of foreign chromosome becomes incorporated into the host-cell chromosome, which is then replicated in mitosis (Chapter 2) each time the host cells divide. Obviously, a transfer of new genetic material into a plant cell can take place in only a very few cells, under special conditions. There is no question, for example, of all the millions of cells in a young seedling having their genetic composition modified by insertion of appropriate pieces of foreign chromosome! The technology of plant genetic engineering actually rests on advances in plant hormone research techniques that have made it possible for plant cells to be grown on artificial media. Furthermore, the full exploitation of recent spectacular developments in molecular biology were made possible not just by the discovery that plant cells could be made to grow and divide in culture but also by the discovery that in the presence of the appropriate plant hormones the growing cells could be made to develop into specific plant organs.

Manipulating cell cultures

Plant tissue culture techniques were developed soon after the discovery of the role of cytokinins as promoters of cell division. An agar medium containing nutrients is placed in a conical flask, plugged with cotton wool and sterilized. On to the surface of this medium is placed a small piece of tissue carefully removed from the central pith of the stem of a tobacco plant.

After a few days the cells begin to divide and enlarge, and this process continues until there is a mass of large spherical cells on the agar surface. Growth and cell division can, in fact, be so prolific that the flask can become almost filled with plant tissue after a few weeks. However, the mass of totally undifferentiated, essentially identical cells will only grow if the ratio of the two growth hormones IAA and kinetin in the agar medium is kept at a specific value.

Further experiments involved changing the amount of cytokinin in the agar medium and transferring on to the new medium a small amount of proliferating tissue from the cell mass described above. Two astonishing discoveries were made. First, if the cytokinin concentration in the medium is reduced to just one-tenth of its former level, growth of undifferentiated tissue ceases and localized growing points develop instead. These emerge as roots, which quickly extend and branch to penetrate the agar medium; no shoots ever appear. Second, if the concentration of kinetin in the agar medium is *increased* there is again a reduction of undifferentiated growth and the development of localized growing points – but in this case they turn green and soon develop all the characteristics of shoots, with buds and leaves. No roots ever develop on this medium!

Depending on the relative amounts of auxin (IAA) and kinetin in the liquid or agar growing medium, plant cells can be made to grow into a mass of undifferentiated callus cells (*top left*), in which there are no recognizable organs, or they can be made to grow roots alone (*bottom left*) or shoots alone (*above*). If young shoots produced in this way are transferred to a root-promoting medium, the development of roots produces a complete plant (*right*) which is perfectly viable and can be potted out in compost and grown on in the normal way.

So, tissue culture techniques developed in plant hormone research have provided the means of controlling absolutely whether plant cells grow into organs or simply proliferate in an undifferentiated manner. Furthermore, if organ differentiation is induced, there is absolute control over whether the organs developed are roots or shoots. The control of differentiation is achieved by the relative balance of IAA and cytokinins in the tissue, a relatively high IAA/cytokinin ratio inducing root initiation, a medium IAA/cytokinin ratio favouring undifferentiated cell proliferation and a relatively low IAA/cytokinin ratio favouring shoot initiation and growth. The balance of IAA to cytokinins therefore appears to control the pattern of development followed by plant cells.

This control of organ development by hormone balance immediately explains the effectiveness of horticultural rooting compounds. Stem cells at the basal end of cuttings are dipped into rooting powder in which the active ingredient is usually indole-3-butyric acid. This is metabolized in the cells to produce IAA and so the relative amount of IAA to the naturally occurring cytokinins in the tissue is increased. This increase triggers root initiation in the basal stem cells – and hence the successful establishment of the cutting.

Returning to the tissue culture techniques and their relevance to plant genetic engineering, it is possible to shake the undifferentiated culture cells gently in a liquid nutrient medium and so obtain a suspension of cells that are more-or-less separated from each other. It is even possible to mount one of these cells in a drop of liquid medium hanging from a thin piece of glass over a depression in a microscope slide. This cell can remain alive and indeed begin to divide in what is known as a hanging drop, single-cell culture. Once the cell has begun to divide it can be placed on an agar nutrient medium in a sterile flask and allowed to continue dividing. If the growing cell mass is then transferred to a medium containing the appropriate ratio of IAA and cytokinins, buds will be produced. These can then be removed and transferred to a root-initiating medium where they will produce roots! After a few months the plantlets can be potted in compost and grown on into fully reproductive plants. A complete plant can thus be grown from a single cell; indeed complete plants have been grown from individual pollen grains. Such plants are, however, sterile; they cannot form gametes (pollen grains and egg cells) because each of their cells is haploid and contains only one set of chromosomes (Chapter 2). It has been possible, however, to treat the cultures of pollen grain cells with substances like colchicine which by some means not yet understood causes the chromosomes to double. The result is that the plants which subsequently develop are fertile and have the added attraction that all the pollen produced will carry only the characteristics in the original set of chromosomes, including recessive ones which may be required in a breeding programme.

The availability of cells separated from each other in a liquid medium opens up the possibility that new genetic material can be implanted into them, after which, following the appropriate treatment, the individual cells can be induced to divide and grow into complete plants in which *every* cell will contain the new genetic material. But there are barriers to the implantation of substantial pieces of large molecules such as DNA, and these barriers are the tough cellulose cell wall and the cell membrane or plasmalemma, inside it.

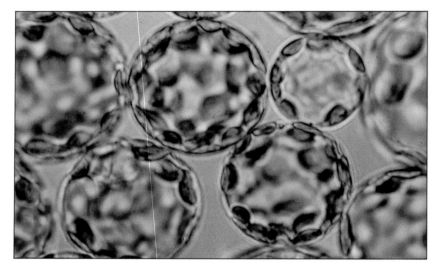

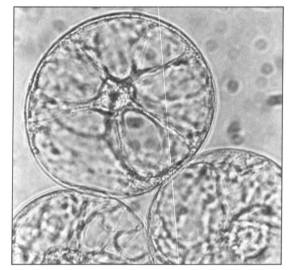

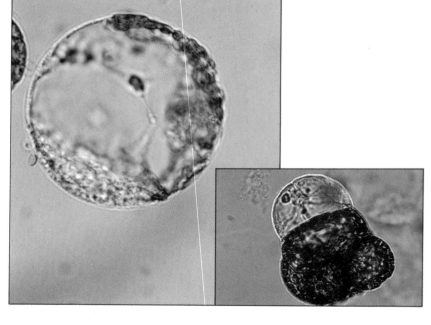

Under special conditions, the walls can be removed from plant cells to leave naked protoplasts enclosed only by cell membranes, rather like animal cells. Naked protoplasts of chlorophyll-containing mesophyll cells were prepared from leaves of *Petunia parodii* (*above*) and from a cell suspension of albino *Petunia hybrida* cells (*left*). A cell from each of these cultures can be seen to have fused (*bottom left*) to form a heterokaryon, or somatic cell hybrid. Sometimes several cells of each type will fuse simultaneously, as shown below where two mesophyll and two cultured cells are seen fusing together.

One method of introducing new genetic material into plant cells is known as somatic cell hybridization. The technique involves taking cultures of the somatic cells – diploid body cells – of two different plants and trying to make the cells fuse with one another. But the cytoplasm of the cells can not fuse unless the cellulose cell walls are first removed because the walls prevent the plasmalemmae and cytoplasm of the two cells from making direct contact. The cellulose walls can be removed by gently treating them with the enzyme cellulase, which dissolves away the wall leaving what is known as a naked protoplast. Fusion between such naked protoplasts has been achieved, and successful hybrid plants have been produced using species of *Petunia* and a few other plants. However, hybridization has proved difficult to achieve with most species, and many difficulties will have to be overcome before this technique becomes useful in the routine transfer of genetic material from one plant to another.

Cell surgery and the biological scalpel

Both natural breeding and somatic cell hybridization depend on the chance incorporation of genetic material into a cell, and then on rigorous selection to identify the hybrid offspring containing the desired characteristics. It would be much better if we could identify just where on a chromosome the gene responsible for a particular characteristic was located, and then be able to cut it out and transfer it to the chromosome of a cell from another plant. So

how can the genes associated with, for example, productivity, disease resistance, susceptibility to cold, or amino acid synthesis be identified and transferred in lengths of chromosome from one plant cell to another?

It turns out that genetic material is actually transferred into plant cells quite naturally by some pathogenic organisms that cause either gross abnormality or disease in plants. The bacterium *Agrobacterium tumifaciens*, for example, causes the growth of tumours and cancers on the sides of plant stems. The disease is called crown gall disease. The bacterium usually lives in the soil, but it can infect plants through wounds where unprotected cells are exposed. The bacterium induces uncontrolled cell division, and what is of particular importance is the fact that the plant's cells continue to divide in an uncontrolled way even when the tumour has been completely cleared of the bacteria. The bacteria must, therefore, have permanently altered the cells in some way so that the capacity of growing and dividing rapidly is transferred to the new daughter cells each time a cell divides. The cells are said to be 'transformed', and they can be cultured in sterile conditions totally free from the bacterium. In addition, unlike normal plant cells in culture, they grow without the medium having to contain IAA and a cytokinin. The transformation of the cells involves the introduction of genes that modify their metabolism so that they produce the appropriate amounts of IAA and cytokinin required for cell division

The tumorous growths on the side of the *Gladiolus* bulb (*below*) and rose stems (*below right*) are known as crown gall disease and are caused by infection of the plants by the bacterium *Agrobacterium tumefaciens*.

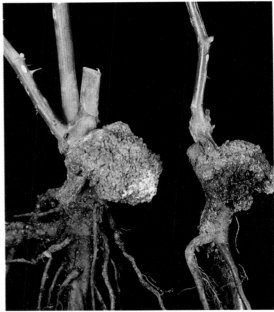

195

and growth. In these tumour cell cultures it is much more difficult than in normal cell cultures to induce the formation of organs such as shoots and roots, although these problems have now largely been overcome by the development of 'disarmed' strains of *Agrobacterium*.

How are these dramatic changes induced by the *Agrobacterium*? They are achieved by the transfer of a 'plasmid' from the bacterial cell to the higher plant cell. Plasmids are rings of double-stranded DNA which are quite separate from the chromosomes. They carry genes which can be transcribed and translated into enzymes in both bacterial cells and those of higher organisms. Some have been studied so thoroughly that their entire sequence of base pairs is known. What happens when an *Agrobacterium* cell comes in contact with a healthy higher plant cell is that the cells bind together and a tumour-inducing (T_i) plasmid is transferred from the bacterial cell to the plant cell. A part of the DNA in the plasmid ring is then incorporated into one or more of the higher plant chromosomes, this incorporated DNA being known as T-DNA. Once incorporated into the plant chromosome the genes on the T-DNA are utilized by the biochemical machinery of the plant cell. The information they contain is read and transcribed into mRNA which is subsequently translated into proteins and enzymes at the ribosomes. Because the T-DNA is incorporated into the chromosomes of the plant cell it will be reproduced and copied each time the cell divides, so that every cell of the tumour will contain an identical length of T-DNA. It is for this reason that the bacteria are no longer required to maintain tumour growth once a cell has been transformed.

The new proteins and enzymes being produced by the presence of the T-DNA in the plant chromosome obviously maintain the tumorous condition of rapid cell proliferation. Several unusual amino acids are also synthesized in the tumour cells; these are called opines, and they are not produced by normal plant cells. They are secreted by the tumour cells, and *Agrobacterium*, but not apparently other bacteria and fungi, can use them as nutrients. This gives *Agrobacterium* a competitive advantage over other bacteria growing on the tumour tissue. In experimentally transformed plant cells the opines also serve the useful purpose of acting as markers, so that these cells can be readily distinguished from those that have not been transformed.

Plasmids clearly provide the means of modifying the genetic information in a plant cell by being able to carry certain genes into the cell and insert them into the plant's chromosomes so that they are replicated every time the cells divide. In theory, therefore, if a method could be found of inserting a particularly desirable gene into the T-DNA portion of the plasmid of a bacterium such as *Agrobacterium*, it would be possible to use the plasmid to introduce the gene into the chromosome of another plant cell where it would become a permanent part of the genetic composition of that plant. This is precisely what genetic engineering is all about. The T-DNA of the plasmid acts as a vector or vehicle with which a particular gene or group of genes can be transferred from one plant to another, thus overcoming all the difficulties of incompatibility which may prevent such gene transfers in natural breeding or somatic cell hybridization.

The gene to be transferred has first to be located in the chromosomes of the donor plant, and then the length of DNA containing it has to be cut out and incorporated into the T-DNA region of a plasmid. This is achieved using one or more special enzymes known as restriction enzymes. These enzymes have the capacity to break the DNA molecules at specific and recognizable places characterized by a special sequence of the component molecules of DNA called bases. The plasmid ring of DNA is then also broken open in the T-DNA region with the same restriction enzyme that was used to break the DNA molecule in the donor plant. The broken plasmid and the fragment of DNA containing the gene to be transferred are then mixed together in the presence of an enzyme called ligase which joins up the broken ends of the DNA molecules so that the short length of plant DNA containing the gene to be transferred is incorporated into the plasmid, whose molecule of DNA re-forms into a complete circle. The plasmid can then be transferred to a totally different plant where the T-DNA portion is incorporated into the host cell's chromosomes.

This procedure has been used successfully in a number of gene transfers. For example, the gene required to make the storage protein phaseolin in French beans has been transferred via an *Agrobacterium* plasmid to sunflower plants. The resulting tumour cells in the sunflowers produced this characteristic protein, which they would normally never manufacture.

The sequential diagram opposite outlines the main procedures of plant genetic engineering. The desired gene is cut out of the chromosome of the donor plant and inserted into the plasmid ring of the bacterium *Agrobacterium tumefaciens*, which will act as the carrier. The bacterium, containing its modified plasmid, is then allowed to 'infect' cultured cells of the plant that is to receive the new gene. The altered section of the plasmid ring becomes incorporated into the chromosomes of the recipient plant cells, and when these are grown on in specially controlled cultures they produce plantlets that have the desired new characteristic.

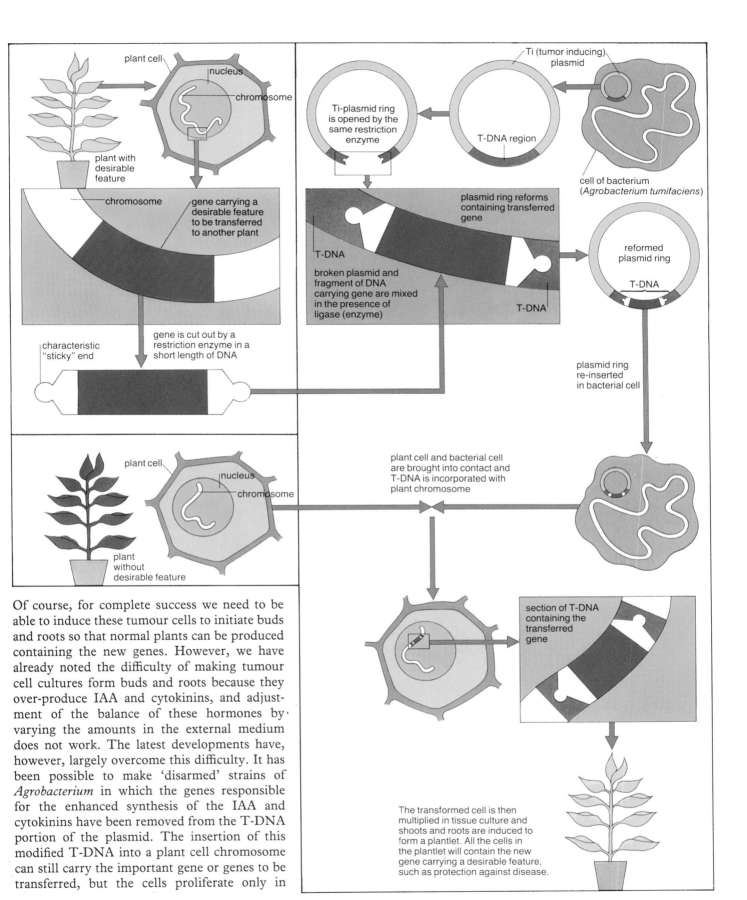

plant cell
nucleus
chromosome

plant with desirable feature

chromosome

gene carrying a desirable feature to be transferred to another plant

characteristic "sticky" end

gene is cut out by a restriction enzyme in a short length of DNA

plant cell
nucleus
chromosome

plant without desirable feature

Ti (tumor inducing) plasmid

Ti-plasmid ring is opened by the same restriction enzyme

T-DNA region

cell of bacterium (*Agrobacterium tumifaciens*)

T-DNA

broken plasmid and fragment of DNA carrying gene are mixed in the presence of ligase (enzyme)

plasmid ring reforms containing transferred gene

T-DNA

reformed plasmid ring

T-DNA

plasmid ring re-inserted in bacterial cell

plant cell and bacterial cell are brought into contact and T-DNA is incorporated with plant chromosome

section of T-DNA containing the transferred gene

The transformed cell is then multiplied in tissue culture and shoots and roots are induced to form a plantlet. All the cells in the plantlet will contain the new gene carrying a desirable feature, such as protection against disease.

Of course, for complete success we need to be able to induce these tumour cells to initiate buds and roots so that normal plants can be produced containing the new genes. However, we have already noted the difficulty of making tumour cell cultures form buds and roots because they over-produce IAA and cytokinins, and adjustment of the balance of these hormones by varying the amounts in the external medium does not work. The latest developments have, however, largely overcome this difficulty. It has been possible to make 'disarmed' strains of *Agrobacterium* in which the genes responsible for the enhanced synthesis of the IAA and cytokinins have been removed from the T-DNA portion of the plasmid. The insertion of this modified T-DNA into a plant cell chromosome can still carry the important gene or genes to be transferred, but the cells proliferate only in

This *Nicotiana* plant has a hidden weapon. It has been transformed by the introduction, from the cow-pea plant, of a gene that enables it to manufacture an enzyme inhibitor that kills insects.

In the years ahead, gene transfers of this kind could play a major role in protecting plants from the pests and diseases that cause so much damage to crops and to stored food. Imagine the impact they could have in the Third World where 70 million tonnes of maize, rice and other cereals are lost every year to insects, fungi and other pests.

One of the ultimate goals of genetic engineering in plants is to create new cereals *opposite* capable of forming symbiotic relationships with organisms such as *Frankia* and *Rhizobium*. This would enable the cereals to develop nitrogen-fixing root nodules *inset* like those that make the legumes so productive, and would transform food production in parts of the world where soils are poor and where artificial nitrogen fertilizers are beyond the means of most farmers.

response to the plant hormone concentrations in the agar medium, and their development into roots and shoots is therefore completely under experimental control.

While the use of plasmids has been spectacularly successful in transforming plant cells, several other methods of gene transfer are now being investigated. One utilizes plant viruses, of which there are two types – the first consisting of a double strand of DNA surrounded by a protein capsule; the other consisting of RNA in a similar capsule. Viruses can not reproduce themselves outside the living cell of another organism which has the biochemical machinery to read and utilize the information contained in their DNA and RNA. Viruses therefore offer the possibility of introducing into plant cells nucleic acid molecules which carry specific genetic information. The cauliflower mosaic virus may have potential as a gene vector though there is a great deal of work to be done before this is likely to become a reality. This DNA virus is very infectious: simply rubbing it on to a leaf will cause an infection that quickly spreads to almost every cell in the plant. The DNA does not, however, become incorporated into the chromosomes of the host cells, but in plants that can be propagated vegetatively this may not be a major disadvantage.

Another method involves the use of 'transpos-

able elements'. These are limited lengths of the DNA molecule which seem to have the capacity to move from one chromosome to another and within one chromosome; they are sometimes referred to as 'jumping genes' because they contain genetic information. They occur in maize and other plants and they may offer another method of transferring genetic material from the chromosomes of one plant to those of another.

Finally, it may be possible to insert DNA molecules directly into plant cells by injection or incubating them in a solution containing the molecules together with other necessary substances. This procedure has met with some success in animal cells, but transfer of DNA into plant protoplast cultures by this method has not yet been successful.

There is no doubt that the way is clear for very major advances to be made in improving the capacity of plants to grow and develop, to withstand disease, drought and cold, to fix nitrogen and to synthesize amino acids and proteins which are essential in a human diet. The way is also clear to make particular crop plants resistant to herbicides, which would greatly assist in increasing yield, and of increasing the efficiency with which they capture and store the sun's radiant energy. But much basic work has to be done in identifying and locating the genes involved in these processes so that they can be effectively manipulated and transferred between plants. There is also much more basic research to be done in developing tissue culture techniques and inducing organ development in monocotyledonous plants, especially in the grasses and cereals which have lagged far behind the techniques developed for dicotyledonous plants.

The challenge of the future is there for all to see. It is now within the grasp of plant biologists to redesign and modify crop plants to satisfy the future nutritional needs of mankind, and to induce plants to produce new compounds with pharmacological properties that will help to relieve suffering. Progress is unlikely to be spectacular, but gradual and based on careful and time-consuming studies of the fundamental mechanisms controlling plant growth, development and metabolism at the biochemical and physiological levels. It is doubtful whether there has ever been such an opportunity to relieve hunger, starvation and malnutrition in the human race, or to do so much damage if it all goes wrong.

Further Reading

Anatomy of Flowering Plants, P. Rudall. Edward Arnold (Publishers) Ltd., London, 1987

About Plants – topics in plant biology, F. C. Steward with A. D. Krikorian and R. D. Holston. Addison-Wesley, Reading, Mass., USA, 1966

Biology of Plants, P. H. Raven and R. F. Evert, Worth Publishing Inc., New York, 1981

Biological Time-keeping, J. Brady. Camb. Univ. Press, Cambridge, UK, 1962

Biological Clocks, J. Brady. Edward Arnold, London, 1979

Lichens – an illustrated guide, F. Dobson. The Richmond Publn. Co. Ltd., 1981

The Life of the Green Plant, A. W. Galston, P. J. Davies & R. L. Satter. Prentice Hall, Englewood Cliffs, New Jersey, USA, 1980

The Living Plant, A. J. Brook. Edinburgh Univ. Press, Edinburgh, 1964

Plant Cell and Tissue Culture, J. Reinert & M. M. Yeoman. Springer-Verlag, Heidelberg, 1982

Plant Physiology, F. B. Salisbury & C. W. Ross. Wadsworth Publn. Co. Inc., Belmont, Calif., USA, 1980

Plant Tissue Culture, D. N. Butcher & D. S. Ingram. Edward Arnold (Publishers) Ltd., London, 1976

Plants at Work, F. C. Steward. Addison-Wesley Publn., Co. NY, 1967

The Plant Cell, W. A. Jenson. Macmillan, London, 1969

Seaweeds and their uses, V. J. Chapman. Methuen & Co. Ltd., London, 1970

The Structure and Life of the Bryophytes, E. V. Watson, Hutchinson University Library, London, 1967

Water and Plants, H. Meidner & D. W. Sheriff. Blackie, Glasgow & London, 1976

Acknowledgements

I am most grateful to many colleagues and friends for assistance in producing this book. In particular, I should like to thank my former student, Dr. Clare M. Anderson, Girton College, Cambridge, UK, for many helpful discussions, allowing me to use original data from her doctoral thesis, and especially for personally donating the blood used in the experiment illustrated on page 60; Mr. Norman Tait for generously giving his professional advice on photography; Mrs. Pamela McEwan for typing the manuscript; Mr. Eric Curtis, Curator of the Glasgow Botanic Garden, for providing many specimens for photography; The Hunterian Museum of Glasgow University for the specimens of fossil plants; Dr. A. M. M. Berrie for providing seeds and identification of certain specimens; Dr. D. D. Clarke and Dr. C. T. Wheeler for suggesting illustrations for Chapter 17; the Department of Botany of Glasgow University for access to microscopic slides from the Departmental Collection, and to Professor R. E. Cleland, Botany Department, University of Washington, Seattle, USA, for permission to photograph the Instron stress-analyser (p168) in his laboratory.

The copyright of the photographs appearing in this book is held by the author except in the case of the photographs listed below for the use of which the author is greatly indebted to the following copyright holders.

Dr. C. M. Anderson, Girton College, Cambridge UK (data on figures on pp 150, 151); Heather Angel (p 57); G. I. Bernard, Oxford Scientific Films (pp 7, 20 top, 21 bottom right, 158, 182 bottom); Dr. B. G. Bowes, Botany Dept. Glasgow University, UK (pp 37 bottom, 165 top left and right, 170 left, 192 bottom, 193 bottom); Prof. D. Boulter, Dr. A. M. R. Gatehouse and Dr. V. Hilder, Botany Dept. Durham University, UK (p 198); J. A. L. Cooke, Oxford Scientific Films (p 37 right); Dr. J. H. Dickson, Botany Dept. Glasgow University, UK (p 41 scanning electron microscope pictures of pollen grains); Michael Fogden, Oxford Scientific Films (pp 87, 180); Prof. J. Woodland Hastings, Biological Laboratories, Harvard University, Cambridge, Mass. USA (p 147 top left and right); Insight Surveys Ltd. (p 62 left and bottom); Prof. T.-H. Iversen, Botany Dept. Trondheim University, Norway (pp 173 bottom, 192 top, 193 top); Rodger Jackman, Oxford Scientific Films (p 145 bottom); Dr. C. C. Jeffree, Oxford Scientific Films (p 118); Dr. B. E. Juniper, Botany Dept. Oxford University, UK (pp 10, 99 bottom, 178 left); Breck P. Kent, Oxford Scientific Films (pp 103, 184 bottom); Zig Leszczyski, Oxford Scientific Films (p 152); Ministry of Agriculture, Fisheries and Food, London (pp 187 top and bottom, 195); Ian Moar, Oxford Scientific Films (p 143); Prof. R. Moore, Biology Dept. Baylor University, Waco, Texas, USA (p 69); Stan Osolinski, Oxford Scientific Films (p 88 top left); Richard Packwood, Oxford Scientific Films (pp 38 top right, 111); Prof. G. Perbal, Laboratoire de Cytologie et Morphogenèse Végétales, Université Pierre et Marie Curie, Paris, France (p 66); Ivan Polurin, Natural History Photographic Agency (p 65); Dr. J. B. Power and Prof. E. C. Cocking, FRS Botany Dept. Nottingham University, UK (p 194); Prof. R. D. Preston, FRS, Biophysics Dept. Leeds University, UK) (p 168 bottom left and right); Rothamsted Experimental Station (p 119); Philip Sharpe, Oxford Scientific Films (p 188); Alastair Shay, Oxford Scientific Films (p 177); Mr. T. Norman Tait (pp 16 left, 19 top, 21 left, 22, 24 top right, 25, 27 bottom left, 30 top and bottom, 36, 38 top left and bottom left, 48, 57 inset, 114, 123 top and bottom, 124 centre and right, 126, 127 top and bottom, 129, 140, 142 top and centre. 147 bottom, 156 bottom left, 175 bottom right, 179, 181 top, 183, 185); Kim Taylor (p 74); Prof. D. Volkmann, Botany Dept. University of Bonn, Federal Republic of Germany (pp 11 bottom right, 92, 133); A. G. Wells, Oxford Scientific Films (pp 142 bottom, 143); Dr. G. C. Whitelam, Botany Dept. Leicester University, UK (84 bottom); Prof. K. V. Wood, Biology Dept. University of California at San Diego, USA (p 189).

Illustrations by Rodney Paull, Eitetsu Nozawa and Ray Burrows.

Index